The Coastal Zone

The Coastal Zone

PAST, PRESENT, AND FUTURE

F. John Vernberg and
Winona B. Vernberg

University of South Carolina Press

UNIVERSITY OF SOUTH CAROLINA **BICENTENNIAL**

© 2001 University of South Carolina

Published in Columbia, South Carolina, by the
University of South Carolina Press

Manufactured in the United States of America

05 04 03 02 01   5 4 3 2 1

Library of Congress Cataloging-in-Publication Data

Vernberg, F. John, 1925–
    The coastal zone : past, present, and future / F. John Vernberg and
Winona B. Vernberg.
        p. cm.
    Includes bibliographical references.
    ISBN 1-57003-394-3
    1. Coastal ecology. 2. Coastal zone management. I. Vernberg, Winona
B., 1924– II. Title.
    QH541.5.C65 V47 2001
    577.5'1—dc21   00-011819

Contents

Tables

Figures

Plates

Preface

Concerns about the health of the environment have become firmly embedded in the public's list of vital issues confronting present and future generations. The emergence of environmental issues as a significant part of the societal agenda can be traced to the 1960s and 1970s, although threats to the world ecosystem were identified much earlier. Along with increased public attention on environmental issues, the informed public also needs a more in-depth understanding of the principles of environmental science, especially as they pertain to the coastal zone.

This book is an introduction for a general readership to become better informed and active in their own coastal communities. This readership ranges from private citizens to high school and college students who are not necessarily majoring in environmental studies.

The chapters in this book are organized to permit the reader to first gain an overview of what the coastal zone is and understand its historical and present-day uses. Chapters 2 and 3 introduce some ecological characteristics of the major coastal zone habitats and the diverse and dynamic environmental processes influencing the coastal region. Building on this background in ecological principles and terminology, chapters 4 through 7 deal with major coastal environmental issues: urbanization, chemical and biological contamination, and other human interventions. Chapter 8 presents an overview of various policies, regulations, and laws pertaining to the management of the coastal zone environment. The final chapter offers some views on the future of the coast.

Within the text of each chapter, we have cited various references. Some document original data sources, while others provide citations for those readers who want more information on a given topic than we can include in this book. A number in parenthesis in the text corresponds to a number in the reference list at the end of each chapter.

The many colleagues and friends who helped with constructive criticism are now formally and gratefully acknowledged. We are particularly indebted to Ms. Lucy Hollingsworth for her invaluable help in preparing the illustrations and Dr. Wayne Beam, who contributed Chapter 8. Karen Pendleton was primarily responsible for preparing the manuscript for submission to the University of South Carolina Press and remained cheerful and uncomplaining despite manuscript changes and rewrites. A number of people critically read either all or part of the chapters and made constructive suggestions: Drs. S.

Stancyk, R. Feller, T. Chandler, G. Scott, T. Siewicki, G. Kleppel, R. Beekman, and A. Lewitus. Dr. R. Zingmark provided several photographs of tropical habitats.

We express our appreciation to the following publishers for the use of figures and tables from their publications: figure 1.2b, Prentice Hall; figure 2.2b, Edward Arnold Publishers; figures 2.3 and 2.5, table 2.1, Springer-Verlag New York, Inc.; figure 2.9, Kluwer Academic Publishers; table 2.4, John Wiley and Sons, Inc.; Figure 3.1, Geological Society of America; figures 4.2, 4.3, 4.6, 4.10, 4.11, 5.2, 5.3, 5.6, 5.7, 5.8, 5.10, 5.11, 6.3, and tables 5.1 and 5.2, University of South Carolina Press; figures 4.4 and 7.3, Ecological Society of America; figure 5.1, Estuarine Research Federation; figure 5.4, American Association for the Advancement of Science; figure 5.5, Marine Technology Society; figure 5.9, CRC Press LLC; figure 6.1, Woods Hole Oceanographic Institution; figures 7.1, 7.2, and 7.4, Duke University Press.

We gratefully acknowledge a grant from the South Carolina Sea Grant Program to partially support publication costs. Many of the studies that illustrate research ion southeastern coastal environmental problems were funded by grants under the general title of Urbanization and Southeastern Estuarine Systems (USES) from the National Oceanographic and Atmospheric Administration.

Last, we dedicate this book to our youngest set of grandchildren, Andrew and Haley Beekman and Stefan Vernberg. Hopefully the coastal environment will persist for the continued enjoyment of this new generation.

The Coastal Zone

1

The Coastal Zone: History and Use

During the geological evolution of our planet, the interface between land and large bodies of water has become a dynamic environment of unique importance not only to a vast array of plants and animals, but also to the development of human society. This complex region where land, air, and water interact exists in two major geographic areas: where large freshwater lakes are located, and where the land meets the sea. Many of the problems and principles presented in this book are applicable to both regions, but it is the melding of sea and land that will be emphasized.

One purpose in writing this book is to provide an overview of the ecological characteristics of the coastal zone that will provide the reader with the background to become aware of environmental consequences resulting from human perturbations, such as chemical and biological contamination and habitat alterations. Competing uses of limited coastal zone resources will be explored, followed by an overview of the present management policies, laws, and regulations pertaining to the coastal zone. A concluding chapter will present our thoughts on the future of the coastal region.

The coastal regions of the world differ in many geological and biological features and in their degree of human development; however, they share many basically similar ecological and economic characteristics. The differences in coastal regions will be highlighted as well as these similarities. Although issues pertinent to many diverse coastal habitats will be presented, we draw more examples to illustrate specific coastal problems from situations existing in the United States, with emphasis on the Atlantic and Gulf Coasts. We will also provide a reading list for those readers who seek more details about a specific region.

Historical overview

Life may have originated in the primordial ooze located between the high and low tides or in adjacent shallow waters. From this ancestral home, various forms of life evolved. As organisms invaded both the adjacent aquatic and terrestrial habitats, they developed mechanisms either for breathing air or for obtaining oxygen from freshwater. Not only have many profound environmental and evolutionary events taken place during the millions of years since life first appeared, but the coastal region has assumed an extremely important role in the history and development of human society (1–4).

Extensive archaeological evidence of ancient human settlements along the coasts of continents documents the early and ever-increasing utilization of resources found there. Further, it is clear that human habitation flourished along the coast of the United States long before Europeans arrived. Although humans occupied localized regions of the coastal zone for fishing, farming, and hunting, the enormity of the adjacent ocean hindered inter-continental exploration and migration until suitable modes of transportation were developed. As improved means of traversing the ocean evolved, coastal settlements grew larger and the surrounding land was increasingly used for terrestrial and water-associated commerce. Harbors provided protection for ships from oceanic waves and storms.

Some early coastal cities still exist today, while others have faded into history. The well-preserved ruins of Ephesus, an early Christian city of mag-nificent buildings, and a major center of culture and trade remain on the coast of Turkey. The location of Ephesus on a harbor connected to the Mediterranean Sea made it a major seaport city. For a number of reasons, the city declined in importance and was finally abandoned. One major change that undoubtedly hastened its demise was the silting in of the harbor, mak-ing it impossible for ships to dock near the city and unload cargo or load cargo for export. Although attempts to maintain the waterways by dredging were partially successful, the increased runoff of silt, resulting from upland development, was too great to maintain a viable port. Today the sea is ten to fifteen miles away from the former waterfront area, which was located in the immediate vicinity of the city.

With the increased technology associated with ocean exploration and the success of explorers in reaching distant lands for trade and settlement, cities strategically located on excellent harbors became increasingly impor-tant. The fall of Constantinople in 1453 to Mohammed II greatly stimulated the development of ocean travel for commercial purposes of trade. One con-sequence of this Islamic military victory was to effectively reduce the impor-tance of the strategically located land-based trade route between the East and the western Christian countries. Since land routes were no longer safe for travel to seek the riches of the Far East, countries sent explorers to find access by way of the sea. A few examples of these daring expeditions will illus-trate the impact of discovering new sea lanes. Dias rounded Africa's Cape of Good Hope in 1486–1488 and Vasco da Gama explored the coast of India in 1497–1499, opening the route to the East Indies for Portugal and initiating a period of prosperity for that country. The expeditions of Columbus to find a western oceanic route to the Far East, which occurred during this same period, resulted in the European colonization of the American continents.

The circumnavigation of the world by the Magellan expedition (1519–1522), sponsored by Spain, further expanded the knowledge of foreign lands leading to trade and cultural exchanges between different peoples. Though Magellan was killed in the Philippine Islands the expedition under the direction of Cano completed the round trip to Spain. Many other daring captains led expeditions to other parts of the world. As a result of this remarkable age of exploration, small countries such as the Netherlands, Great Britain, and Portugal became international powers—due in large part to their leadership in exploration via the sea and their development of aggressive trade policies. The economic success of many present-day countries continues to be closely tied to ocean shipping and the expansion and upgrading of coastal facilities.

Today the coastal region is being subjected to intense multiple stresses resulting from conflicting demands for use of finite resources by an ever-increasing population. For example, about 50% of the people living in the United States are located in coastal areas, which constitute less than 10% of the land area of the forty-eight contiguous states. Toward the middle of this century the population center of the United States began shifting to coastal regions. Demographic predictions indicate that people will continue to migrate to the coastal regions, and by early in the twenty-first century more than 70% of the population will be located in the coastal regions of marine waters and the Great Lakes.

Serious problems already exist that are associated with the conflicting demands on a limited resource base resulting from high population densities. To utilize these limited resources to the maximum benefit of present and future generations requires active participation of the populace in determining resource management policies based on understanding the interaction between ecological, economic, and social factors. Rather than continuing our present policy of permitting a series of piecemeal, immediate, short-term development projects, long-term integrative planning with active participation by an informed public is vitally needed.

What is the coastal zone?

In general the coastal zone encompasses both the neighboring uplands and the adjacent salt waters that are mutually influenced by the interactive complex of various ecological processes (natural and human-influenced) occurring in each region (figures 1 and 2). Various problems are associated with establishing both the landward and the seaward boundary lines. Scientifically, variation in such features as habitat, soil type, and land formation complicate the establishment of a boundary line of the coastal zone. Politically, a legislative and regulatory definition of the coastal zone (which is not neces-

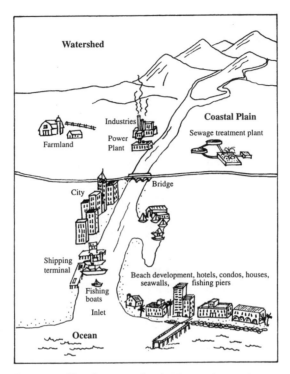

Figure 1.1. **A generalized watershed.** Water drains from mountains via the coastal plain and an estuary to the sea. Some of the human activities influencing water flow and water quality are indicated.

sarily based on sound environmental grounds) has led to conflict among different levels of governmental and regulatory agencies (5–7). For example, all lands seaward of a distinctive geographic point, such as a highway or a county line, may be designated as the landward boundary of the coastal zone. But this artificial boundary may extend through the middle of an extensive intertidal wetland which, from an environmental viewpoint, would be considered a vital part of the coastal zone ecosystem. Although it is difficult to define these boundaries, some information exists that is helpful in considering the need for such a definition.

The seaward boundary can vary depending on a number of factors. For example, in a region where a major freshwater system like the Mississippi River empties into the ocean, the impact of land-based activities will be evident farther from land than in a localized region with limited freshwater input. Also, waters near large coastal population centers are subjected to a wider range of environmental stresses than those adjacent to more pristine uplands. A scientific definition can be based on chemical and physical char-

Figure 1.2a. **Generalized representation of coastal wetland habitats.** The uplands are dominated by the maritime forest (plants near the marsh are influenced by wind and salt spray, while species further inland are less salt-tolerant). The salt marsh community, extending seaward, consists of several salt-resistant species but is dominated by *Spartina* grass. See Barry (12) for detailed account of species composition inhabiting these habitat types, especially those found in South Carolina.

Figure 1.2b. **Generalized representation of coastal rocky headland habitats.** Unlike the Atlantic and Gulf coasts, which have gently sloping seashores, along much of the Pacific coast the mountains extend relatively close to the sea, resulting in steep, coastal rocky slopes. In some areas the mountains level out to form marine terraces bordered seaward by rocky slopes. At the base of these slopes exists a rocky intertidal zone. In regions where streams and rivers enter the sea, beaches and wetlands may be found (adapted from Klee [10]).

acteristics of seawater. Open oceanic water is chemically different from coastal waters, and the physical dynamics of water circulation may vary between these two regions. But differences in chemical or physical characteristics do not result in a fixed line of demarcation, since seasonal changes in the patterns of oceanic circulation and chemical mixing can change the exact geographic location of the boundary line. In general, coastal waters are defined as the water mass extending seaward to the edge of the continental shelf, generally to a depth of about 200 meters.

An intertidal zone marks the boundary line between coastal waters and the uplands (see chapter 2 for discussion of the intertidal zones and tides). However, as is the case with defining any of the boundaries, the line between the uplands and the high tide mark may be defined differently by various regulatory agencies. The high tide mark varies with weather conditions and type of tide. At the time of spring tides, seawater extends to both higher and lower intertidal elevations than during neap tides, when the difference between high and low tide levels are less extreme. Because of this variation, some scientists use the mean high tide to demark the boundary between the uplands and the aquatic wetlands rather than rely on a fluctuating high or low tide mark.

The landward limit of the coastal zone is equally difficult to define. The effects of the ocean may be observed a great distance inland. For example, in some coastal regions salt water penetrates river systems much farther inland than in other geographical areas. As a result there is greater interaction between marine waters and upland habitats located farther from the sea. By contrast, coastal uplands devoid of significant riverine systems tend to be a relatively narrow band of land primarily influenced by onshore oceanic winds. In some coastal areas, high mountains are immediately adjacent to coastal waters, resulting in little or no flat land being present, whereas other regions devoid of mountains may have extensive plains bordering the ocean. Thus the width of upland considered to be part of the coastal zone may vary from a few feet to hundreds of miles.

In summary, although there are difficulties in establishing fixed geographic boundary lines of the coastal zone, it is important for scientific, legal, and resource management purposes to define and understand the extent of the coastal zone. Furthermore, we should be constantly aware that the various coastal zone habitats do not exist in a vacuum, but they are an integral part of planet Earth and are subject to the influence of ecological changes in other neighboring habitats. Pollutants introduced into the air from a smokestack industry hundreds of miles from the coast or chemical run-off from inland sources, for instance, can be ultimately introduced into the coastal zone environment (figure 1).

The coastal area is also influenced by events originating great distances away and not caused by human interventions. For example, hurricanes and tsunamis have their genesis at sea but can destroy a coastal area. The impact of hurricanes along the Southeastern and Gulf coasts of the United States extends from destruction of ships caught in the path of a hurricane, to destruction of buildings on land, to massive flooding and to the death of many residents. Not only do these storms dramatically influence the present-day coastal environment, but they have had a profound impact on the history of early America (see chapter 7).

A holistic view of the complex, interrelated nature of a coastal environment that is subjected to both "natural" short-term and long-term changes and to impacts of human interventions must be an essential part of the decision-making process in resource management.

Uses of the coastal zone

Although some coastal areas have always been attractive as sites for population centers, others were less desirable and more difficult to settle. In the Southeastern United States and as far north as Philadelphia and New York, some wetland areas were considered to be unhealthy by early settlers. Yellow fever and malaria were prevalent in coastal communities all along the eastern sea coast in the 1700s and 1800s. Both diseases are transmitted by species of mosquitoes that flourish in coastal wetlands. In addition, ships coming in from all over the world brought in a continuous supply of previously infected immigrants. Eventually, through a combination of mosquito control measures and quarantine of infected individuals, coastal areas became less deleterious to humans and subsequently a more desirable environment for settlement.

In earlier years, coastal zone resources were considered to be unlimited, and there was little—if any—environmental concern for their use. Now, however, a general awareness exists that coastal resources are finite and that programs of regulation and planning are needed to wisely use these resources for enhancing human society and sustaining the earth's ecosystems. With the anticipated increased population in the coastal zone, the competition for use of resources will become ever more intense, and the potential for irreversible damage and change to this ecosystem will be greatly increased. Thus, it is vital that the present-day populace have a broad overview of the complexities of this important segment of the earth. Further, citizens should become involved in the process of formulating land use and regulatory policies that provide the basis for wise and sustained utilization of the coastal zone.

For overview purposes, the major uses of the coastal zone can be artificially divided into three categories: upland uses, intertidal and shallow coastal water uses, and offshore coastal water uses. Obviously overlap can occur, as well as multiple uses involving activities in two or three of these categories. For example, offshore drilling activity requires an upland-based site for operational support as well as offshore facilities.

Upland uses

With the rise of cities during ancient times and continuing into the present, many of the larger population centers in the world are or were located in the coastal zone. Even today, eight of the ten most populous cities in the world are located in the coastal zone (8). These cities serve as centers for both land-based and oceanic commercial activities. Associated with these cities are large residential areas, industrial sites, commercial buildings, various types of industrial facilities, ports, transportation facilities (including trains, trucks, and automobiles), and infrastructure such as schools, hospitals, sewage treatment plants, and waste disposal sites. Outside of city boundaries numerous demands are made on the coastal zones for vacation and retirement homes, farms, forests, and recreational facilities (marinas, amusement parks, camping areas, and nature parks). In addition to large cities, many smaller cities, towns, and villages exist that arose to meet special needs associated with specific uses. Some older towns were founded around harbors, which were used for small commercial and recreational fishing activities. Others served as centers for agricultural and forestry pursuits.

One of the major uses of the coastal zone in the United States in the past thirty years has been the increased development of facilities for recreational and retirement purposes. This is especially true in the southeastern region of the United States, which has more undeveloped lands with fewer large coastal population centers compared to other regions of the country. The extensive development of the coastal zone of Florida is an example of the trend for increased human activities utilizing limited resources. Other less-developed southeastern states now are experiencing the impact of a similar demand for recreation and retirement living.

The continued increase in human activities in the coastal zone has caused a shift in use patterns to meet the demands of new industries and urbanization in addition to the pressing demands for recreation and retirement facilities. This trend can be illustrated by the expanded urbanization of the greater New York City region. From a small original settlement, the population, commerce, and industry expanded to adjacent areas. To enable the city to grow, upland forest and agricultural lands were converted to urban uses, and many intertidal and shallow water habitats were destroyed by dredging and filling activities. New York's oyster industry provides a well-documented case history of the impact of changes in adjacent waters over the past 150 years (9). In the early 1800s extensive oyster beds were found throughout much of the lower Hudson River estuary, extending from Sandy Hook, New Jersey to Ossining, New York. Productive oyster grounds also existed in the upper harbor. However, by the 1870s polluted shellfish beds from Staten Island, Newark Bay, and the Hudson River resulted in unmarketable oysters. Pollution from solid waste, coal, oil, heavy metals, and upland runoff contributed to this problem. By 1905, the discharge of untreated sewage contributed to eutrophication and was also linked to typhoid outbreaks that were partly attributed to the consumption of raw oysters. By 1921 all commercial shellfish beds in the lower harbor were condemned for public health reasons. Presently, through dredging and filling, most of these areas have disappeared entirely.

One result of the increasing demands for multiple uses of the coastal zone is that vast areas of intertidal zones and wetlands have been altered to provide more uplands for uses requiring dry land. Concurrently, dredging of the intertidal/shallow water region has occurred both to provide increased accessibility for boats and for fill to create highland. This activity has been extensive. Since the mid-1950s, it is estimated that the United States has lost about 450,000 acres of saltwater and freshwater wetlands every year, an area about half the size of Rhode Island. Although at present saltwater wetlands are better protected legally and the rate of their destruction has been slowed, modification of the landscape in the coastal zone needs to be carefully monitored. Changes in the uplands may have a negative impact on wetlands by altering surface water runoff and water basin drainage patterns.

In earlier times agriculture and forestry were important uses of the uplands. But as the coastal population increased, many acres of natural forested areas and farmland were destroyed to accommodate the needs associated with shipbuilding, industrialization, urbanization, and development of recreational/retirement facilities. In those areas still used for agriculture and forestry, newer farming techniques and the increased use of herbicides and pesticides has had an impact on the water quality of adjacent wetlands and streams.

Intertidal and shallow coastal water uses

Along the coast, indentations in the coastline result when bodies of water become semi-enclosed by land but maintain a connection to the sea (see chapter 3). Frequently these bodies of water are associated with the mouths of river systems. They are known by several names —bays, estuaries, sounds, inlets— depending in part on how they were formed geologically. Here ships

National Estuarine Research Reserve System

designated ■
proposed △

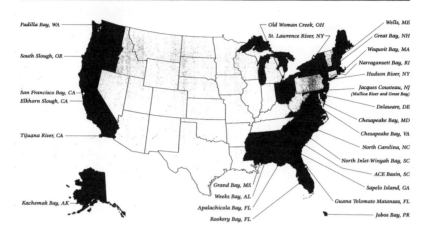

Padilla Bay, WA

South Slough, OR

San Francisco Bay, CA
Elkhorn Slough, CA

Tijuana River, CA

Kachemak Bay, AK

Old Woman Creek, OH
St. Lawrence River, NY

Grand Bay, MS
Weeks Bay, AL
Apalachicola Bay, FL
Rookery Bay, FL

Wells, ME
Great Bay, NH
Waquoit Bay, MA
Narragansett Bay, RI
Hudson River, NY
Jacques Cousteau, NJ
(Mullica River and Great Bay)
Delaware, DE
Chesapeake Bay, MD
Chesapeake Bay, VA
North Carolina, NC
North Inlet-Winyah Bay, SC
ACE Basin, SC
Sapelo Island, GA
Guana Tolomato Matanzas, FL
Jobos Bay, PR

Figure 1.3. **The National Estuarine Research Reserve System** comprises twenty-three designated and five proposed reserves located throughout the nation's coastal zone (NOAA [11]).

can find a safe haven from oceanic storms and unrelenting wave action. Adjacent to both these semi-enclosed waters and to the open ocean, we find an intertidal region that results from the ebb and flow of the tides. Salt marshes are located in intertidal regions of water protected from the eroding effects of wave action. Freshwater marshes are located further up a river system away from the influence of seawater; however, such marshes may be tidally influenced.

The open beach intertidal and shallow-water habitats are primarily used by humans for recreational purposes such as fishing, swimming, and boating. They are also used by commercial and sports fishing to provide various kinds of seafood. In some areas, coastal shallow waters have been used for disposal of societal wastes, including industrial effluents and sewage. To renourish eroding beaches, sand is often dredged from the offshore ocean bottom and deposited on a beach. In other regions, dredged material from semi-enclosed waters (originating both from the intertidal zone and submerged bottom of these waters) is disposed of offshore or on open beaches. This practice most often occurs in places where convenient upland dredge disposal sites are scarce because of extensive upland development and the cost to transport the spoil further inland is prohibitively high.

Some intertidal and coastal water areas have been designated as nature preserves or marine sanctuaries, although these are relatively few in number (10, 11). The National Estuarine Research Reserve Program and the National Marine Sanctuaries Program of the National Oceanographic and Atmospheric Administration (NOAA) are excellent examples of cooperative state/federal programs established to promote education of the public about the value of the coastal and marine environments and to foster research activities addressing coastal zone problems (figure 3). These preserves are generally marine environments that have special ecological qualities and represent unique habitats or are major habitats. Because these sites represent areas that are potentially threatened by human activities, it is important not only to maintain existing sites but to establish additional preserves and sanctuaries to insure that future generations will have the opportunity to see what existed in the past.

To augment the supply of seafood obtained from natural fisheries, there has been an increasing interest in growing marine organisms for human consumption in controlled environments such as ponds or artificially closed-off areas of oceans or estuaries. In some parts of the world, especially in the tropics, ponds are commonly used to raise shrimp. The term *mariculture* refers to the growing of commercially important organisms in seawater, whereas *aquaculture* is a general term applied to the commercial cultivation of either marine or freshwater organisms. Aquaculture has been so successful that today pond culture produces more shrimp for sale than do fishermen harvesting wild populations. In some areas, permits can be obtained from governmental agencies to grow such organisms as clams and oysters in specifically leased parts of the natural habitats.

Submerged lands have been used for mining and extractive industries, including gas and oil production, which are among the most valuable of the various resources found there. Oil and gas fields associated with marine environments tend to be found offshore, whereas some other mining activities are located in salt marshes or shallow-water areas (phosphatic deposits, for example, are typically are found under marshes). Shell, sand, and gravel are also mined from shallow water habitats. These nonliving minerals are considered to be *nonrenewable* resources. By contrast, fishing and mariculture are considered to be living, *renewable* resources, although some fishery species have been harvested so heavily they cannot renew themselves and may become extinct.

Offshore coastal water resources

The value of the vast area of water adjacent to the United States mainland has long been recognized. Important shipping lanes, including those used for national defense and commerce and for transoceanic shipping to local ports exist in coastal waters. However, not until the 1970s were a number of significant laws enacted to protect and manage this region. For centuries the ocean had been considered to be so vast that it was preferred to the valuable uplands for waste disposal, and coastal communities often disposed of wastes into the sea. In 1972, the Marine Protection Research and Sanctuaries Act at last controlled the dumping of wastes into U.S. coastal waters.

Coastal waters also support commercial and sports fishing. To ensure a sustainable harvest of fishery species, the Fishery Conservation and Management Act of 1976 established a fishery conservation zone extending 200 nautical miles from the mainland. Chapter 8 discusses in more detail various state, national, and international laws, conventions and treaties influencing uses of these waters. A limited number of nature preserves and marine sanctuaries have been established in offshore regions, as well, to protect unique natural habitats and special features such as coral reefs and historical shipwrecks. One use of the coastal zone for which a dollar value cannot be assigned is the personal pleasure of viewing the beauty of this region. The exhilaration of observing the wildlife that abounds at the seashore or watching waves pound on an open beach are experiences that cannot be quantified and that have special meaning to each individual. With the sharp shift in population from rural regions to urban settings, many city dwellers seek the quiet, esthetically appealing landscape of the coast. Although esthetic values cannot be directly measured, estimates of the income derived from coastal tourism indicate that tremendous sums are generated. Since a principal reason to be at the coast is to enjoy its beauty, it follows that esthetic delight has a high economic value.

Uses of the coastal zone by nonhuman species

The coastal region is inhabited not only by humans, but by many species of plants and animals. Although it is easy to recognize the economically important species because of their size and abundance, many other plants and animals inhabit the coastal region. The economically important species together with the other biotic members form a complex environment in which the ecosystem's survival is dependent upon dynamic, interspecific relationships. Because the public is unaware of these noncommercial organisms and their ecological importance, their role is not appreciated and they are not protected.

Some plants and animals live in the coastal zone throughout their life cycle, but other organisms only use coastal zone habitats for a portion of their normal life cycle. For example, most offshore species of fish and shrimp spend only part of their life cycle in estuaries for feeding, growth, protection from enemies, and/or reproduction before migrating seaward to the open ocean. Many inland migratory birds stop over in coastal regions to feed during their extensive flights to other destinations. Often human activities have a detrimental effect on the permanent and transitory biota, with the result that the coastal zone can be radically changed and species that appeal to humans disappear or are greatly reduced in number.

The following two chapters will present a broad overview of the ecological principles and characteristics of the various coastal zone habitats, including a discussion of the interactions among the various species and the influence of the nonliving environmental factors such as temperature, salinity, light, and sediment.

In summary, the coastal zone of the world has become increasingly important to human society over time because of population growth. It is a complex region consisting of uplands, various intertidal habitats (including extensive wetlands), harbors, bays, inlets estuaries, open beaches, and inshore coastal waters. During the course of history, it has been used for multiple purposes. Today there is a constant conflict over the use of the finite renewable and non-renewable resources located in the coastal zone. The intensity of conflict among competing users is likely to increase, since most demographic projections indicate that population pressure will grow in the years ahead. A failure to understand the linkages among the environmental, economic, and social implications of any proposed use, coupled with a lack of broadly based land-use planning, is an open invitation to coastal chaos.

2

Ecological Characteristics of the Coastal Zone

Although the earth abounds with diverse types of *biota* (plants and animals) inhabiting many distinctly different regions, life as we know it is restricted to a relatively thin surface surrounding our planet. This limited life-supporting zone is called the *biosphere*. Within the biosphere, many identifiable environmental factors interact differentially, but continuously, to produce a number of distinctive sites called *habitats*. The sum total of the complex interrelationships between these various factors is known as the *environment*. For convenience of discussion, the environment can be divided into two components: biotic, which includes all interactions involving living organisms; and abiotic, which deals with nonliving factors. Biotic and abiotic components influence each other to determine the viability of living matter. Included in the biotic component are all interactions between two or more individuals of the same or different species, as well as the internal physiological processes of an individual. Examples of interactions between individual organisms are competition, predator-prey relationships, and symbiosis. By contrast, individual internal processes encompass such functions as growth, reproduction, neural integration, and metabolism. The abiotic component includes physical, geological, and chemical factors (temperature, salinity, light, tides, nutrients and sediments, to cite a few examples).

Habitats can be described and named for their specific biotic and abiotic characteristics. For example, rocky shores are a type of habitat based on the principal nonliving feature of extensive rocky deposits. By contrast, the salt-marsh habitat is dominated by a living entity, salt-marsh plants. Although a distinctive habitat type exists and can be described in detail, it is not immutable, nor does it exist in a vacuum apart from other kinds of habitats. Environmental changes can alter the characteristics of a habitat to the extent that it is transformed into a different habitat altogether. This transformation may in turn affect adjacent habitats. Such changes demonstrate two important principles: that changes in one part of the biosphere may have profound effects on other parts of the biosphere, and that changes cannot always be predicted with certainty. The magnitude and time period of change may be on a macro-scale having a long-term global influence (global warming or the ice ages, for example), or the change may be more localized and be on a micro-scale of varying temporal periodicity. For example, dredging and filling of a marsh will permanently alter such a habitat in a restricted region, whereas controlled burning to maintain a southern pine forest has a transitory effect and must be repeated to prevent the invasion of other undesired tree species.

In recent years, increased scientific attention has been given to viewing the planet Earth as one large interrelated ecological system exhibiting a complex interrelationship between biotic and abiotic factors. Although in the 1800s scientists were aware of the importance of the impact of nonliving factors on organisms, not until 1935 did a scientist named Tansley propose the term *ecosystem* to emphasize the interaction of the diverse parts of the total environment (1). Traditionally an ecosystem may be defined an ecological unit that includes all of the organisms and their physical environment as well as their interactions necessary to maintain life. With the development of advanced analytical and computer technology, a much more sophisticated approach has developed that has permitted systems ecology to emerge as a dominant subdiscipline of ecology.

For purposes of study, systems ecologists have analyzed systems of different sizes ranging from the entire planet to macro ecosystems, such as a tropical rain forest, to artificial, small-scale, laboratory micro ecosystems. Whatever their sizes, the boundary lines of the systems being examined must be defined to permit comparison of results with those from other studies. Because of the complexity of studying natural systems in the field, some investigators have examined simplified artificial systems under controlled laboratory conditions. Although it is difficult to compare laboratory findings with field data, this process does usually yield meaningful concepts and understanding of how systems function.

Biotic factors
Generalized ecosystem

A simplified ecosystem is graphically represented in figure 1. In this figure the primary source of energy, which is required to run the entire system, is light energy from the sun. Green plants utilize this radiant energy in the process of photosynthesis to produce organic material. A simple representation of this process is that chlorophyll-containing plants utilize carbon dioxide and water in the presence of light to produce glucose and oxygen. Energy from the sun is stored as potential energy in the glucose. In recent years research has demonstrated that some ecosystems, such as the area associated with hydrothermal vents located in the dark, deep, bottom regions of the sea, do not depend on photosynthetic organisms for energy but derive energy

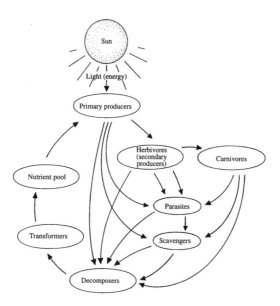

Figure 2.1. **A generalized scheme of an ecosystem.** Arrows indicate flow of energy from one trophic level to another.

instead through a process called *chemosynthesis*. By this process organisms (*chemoautotrophs*) use chemical oxidation of simple inorganic compounds to obtain energy for carbon dioxide conversion into cellular components. They can convert ammonia to nitrite, nitrite to nitrate, sulfide to sulfur, and ferrous to ferric iron. Although these organisms can grow in the dark, they typically require oxygen.

Whether energy is derived from the sun or by chemical transformation, organisms that incorporate the energy into organic matter themselves are called *autotrophs*, while organisms that do not produce their own food are referred to as *heterotrophs*. The output from the autotrophic process, expressed as the creation of new plant tissue, is called *primary production* and constitutes the first nutritional, or trophic, level. The next higher trophic level is reached when autotrophs are consumed by organisms called *herbivores*, who incorporate autotrophic material into new organic compounds. This process is referred to as *secondary production*. *Carnivores* can feed on herbivores or on other carnivores, representing still another level of trophic complexity. *Parasitic organisms* derive their energy from autotrophs, herbivores, carnivores, or even other parasites. Upon the death of an individual of a species, *saprophytic organisms* feed on the remains. The breakdown of complex organic compounds from dead organisms to inorganic material is accomplished by *decomposers*. Certain organisms, called *transformers*, convert inorganic compounds into chemicals that can be used by primary producers. Saprophytes, decomposers, and transformers may be eaten by other organisms.

The food chain and the food web

The pathway of energy transfer from one trophic level to another is referred to as a *food chain* or a *food web* (figure 2). Some food chains are fairly simple: one species feeds only on a certain species and, in turn, is fed upon by another specific species. In this simplified food chain, if one species is not available, the chain is broken and species higher on the chain will starve. Also, if one species near the beginning of the chain is greatly reduced in number, other species higher in the chain may be adversely affected. Organisms at the base of a food chain are often nondescript, minute, and not easily recognized by nonscientists. Hence their fundamental value to the production of organisms of commercial or aesthetic value (shrimp, for instance, or birds) is not always understood or appreciated, and the habitat of the obscure species at the base may not be adequately protected from human perturbations.

Most trophic relationships between organisms are not simple chainlike configurations but are very complex and interwoven; hence, this nexus of interactions is called a food-web (figure 2). Some organisms prey on a number of other species and, in turn, may be devoured by a number of different species. However, others have specific food preferences and feed only on one or a few species. Animals that can feed on either plant or animal material are called *omnivores*, those feeding exclusively on plants are *herbivores*, and those limited to flesh-eating are *carnivores*. Other feeding types (among many too numerous to list here) include detritus eaters or *detritivores* (*detritus*, a major food source in estuarine systems, usually refers to the breakdown of organic particles resulting from the physical breakdown and decomposition of biota) and *planktivores*, which eat plankton (2).

Not only do animals exhibit diversity in the type of food they consume, but they have evolved tremendous variation in the mechanisms they use to obtain food from their environment. Knowledge of this dynamic interaction between food-gathering techniques and how the environment is managed is fundamental to the development of environmental protection policy. For example, oysters, scallops, some fish, and crustaceans feed by filtering small particles from the surrounding water as it passes over specialized anatomical structures of the animal. If the amount of suspended silt is increased too much by such human interventions as dredging or channel construction, the

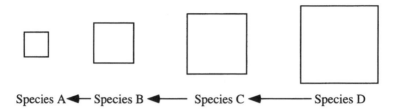

Species A ◄— Species B ◄——— Species C ◄——————— Species D

Figure 2.2a. **Simplified food chain.** Typically larger sized species consume smaller organisms. The food chain is usually limited to four to five links.

filtering mechanisms become clogged and the organism may die. In addition, pollutants dissolved in the water mass may be absorbed, resulting in impaired functioning or death of the organisms. Other species live in the bottom sediments of coastal waters and feed by ingesting large particles or masses of benthic deposits. If these deposits contain toxic substances, the resident organisms may be adversely affected. Thus, the impact of water quality and sediment control activities on the feeding behavior of coastal aquatic organisms must be incorporated into the development of coastal management policies.

Food-web dynamics also control the population size of different coastal species. The interaction between predator and prey is complex and is influenced by many factors. In general, the size of the predator population is dependent on the quantity of available prey; consequently, any fluctuations affecting the prey population will also affect the predator population. Predators typically will consume more prey as prey density increases up to a certain point, then prey consumption will level off. Predators face a dual problem: they must consume enough prey to maintain a population, while not endangering survival by depleting the prey population. An imbalance between these two processes results in marked changes in population size of both prey and predator. Although there are multiple examples of extreme fluctuations in population size, most populations in nature are fairly stable in their density within some range of modest fluctuation. However, there are instances of startlingly explosive growth among animals that have been transplanted to a new ecological situation. A classic example is the great increase in the striped bass population after the species was transplanted from the Atlantic Ocean to the Pacific Ocean. Between 1879 and 1881 a total of 435 striped bass were transplanted from New Jersey to the San Francisco area. Within ten years they were being caught commercially on the west coast,

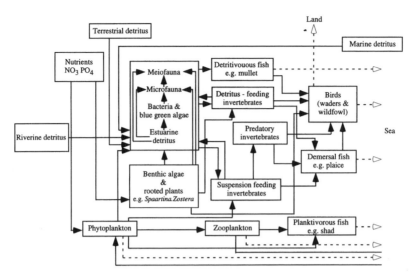

Figure 2.2b. **Generalized estuarine food web.** Dotted lines denote losses from the estuarine system. Arrows indicate flow from one ecosystem component to another (modified from Barnes [22]).

and by 1899 the yearly commercial catch was 1.2 million pounds (29). See chapter 7 for additional examples of introduced species.

In addition to predator-prey interactions, organisms compete for specific environmental resources. Such competition may occur between individuals of the same species (intraspecific competition) or between two or more species (interspecific competition). Organisms actively compete for food, space, and mates, among other essential resources. Competitive success is not solely determined by the internal functional capabilities of an individual, but also by abiotic factors acting directly or indirectly on the organisms involved.

Symbiosis

Not only are trophic interactions between organisms of fundamental importance, but other types of associations among different species of organisms are extremely numerous and varied, ranging from casual contact to absolute physiological interdependence of one or both organisms involved. As a result of the many attempts to describe these various types of associations, a rather voluminous and complex terminology has evolved over the years. Unfortunately, the various terms do not always have the same meaning to everyone.

In the context of this discussion the term *symbiosis* designates the relationship between two or more species living together, and embraces three main categories: mutualism, commensalism, and parasitism. It should be emphasized, however, that many associations do not exactly fit any of these three categories but instead represent intergrades between them.

Mutualism may be defined as a symbiotic relationship that is advantageous to both species involved. In many mutualistic associations, the two species are not together continuously and their dependency may be limited to one factor, as in the relationship between the "cleaner" shrimp and certain reef fishes. The shrimp seem to be responsible for removing the external parasites of the reef fishes, and in turn the shrimp depend on these ectoparasites for their food (3). In other mutualistic associations, all stages in the life cycle of the two species concerned may be almost completely interdependent. Mutualistic associations in which the metabolism of both partners is supplemented are distributed widely in various species. The shipworm *Teredo*, for example, has the ability to destroy wooden piers and harm wooden ships because it houses wood-eating symbionts in its gut. These symbionts transform the ingested wood into energy for the worm, whereas the wood-eating symbionts benefit by being provided a secure habitat and a source of food.

In both *commensalism* and *parasitism*, one of the two species involved is largely dependent upon the other. Commensalism is commonly defined as a relationship between two organisms that is advantageous to only one of the organisms but is not harmful to the host animal. By contrast, parasitism is described as an association in which the unilaterally dependent species is harmful to the host species. As noted by others, however, pathogenicity is not a valid criterion for judging parasitism, as both physiological and environmental conditions may determine whether the dependent organism will have a detrimental effect on the host species. Some organisms classified as parasites will be extremely pathogenic in one host species but relatively harmless in another. On the other hand, some dependent organisms classified as commensal do cause injury to the host, even though these injuries are considered inadvertent. For example, the limbs of the oyster crab *Pinnotheres ostreum*, which lives in the mantle cavity of the American oyster *Crassostrea virginica*, often become entangled in the gills of the oyster so that when the crab tries to free itself, the gills of the host animal are lacerated. Therefore, the oyster crab has been classified as a parasite, although this crab is a filter feeder and does not feed upon the host tissue. Thus, perhaps a more operational definition would describe parasitism as the metabolic dependence of one species on another, and commensalism as a dependence based on utilization of excess food or providing shelter in the burrow or tube of the host animal.

For the purposes of the present overview of the coastal zone, it is important to realize that these various symbiotic and feeding patterns may involve organisms from different habitats. For example, terrestrial organisms may feed on aquatic and/or intertidal species, and coastal water species may feed on intertidal organisms. The interdependence of organisms from these different habitats is another example of how important it is to consider impact of a proposed environmental alteration on adjacent habitats. For example, construction of a retaining wall along the edge of a creek may increase the terrestrial land area, but it will decrease the amount of intertidal zone available for feeding by terrestrial and aquatic species.

Abiotic factors

So far the discussion has focused on biotic components of the ecosystem, but of great importance are abiotic factors, which significantly influence or control biological activities. Temperature, for instance, can determine the ability of an organism to eat, reproduce, or metabolize. The abiotic component includes physical, chemical, and geological characteristics of the system.

Temperature

One of the most extensively studied physical factors is temperature. The massive database on thermal characteristics of the coastal zone reflects both the ease and precision with which it can be measured and the demonstrated profound effect that temperature has on biotic and abiotic processes. Temperature may vary with time at any one location, with latitude, altitude, and with depth of the water. For example, within the temperate intertidal zone, the temperature during the summer may vary daily over a range in excess of 30°C, depending on the amount of cloud cover and the stage of the tidal cycle. By contrast, the deeper waters off the coast experience only slight daily fluctuations (tenths of a degree), but can vary seasonally (several degrees). Geographically, temperature varies with latitude, with little seasonal fluctuation observed in tropical regions, whereas pronounced changes occur in higher latitudes. In transition zones, which occur near the extreme of the distributional range of an organism, unusually high or low temperature may result in mass mortality and extinction of a local population or a species. Figure 3 graphically depicts the relationship of the thermal lethal limits of organisms and their latitudinal distribution. Polar animals are typically not tolerant of temperatures above 8°C, while tropical species usually cannot survive temperatures below 15°C. This phenomenon of mass mortality due to unusual extremes in temperature is not limited to populations found at their distributional limits but can be observed throughout their distributional

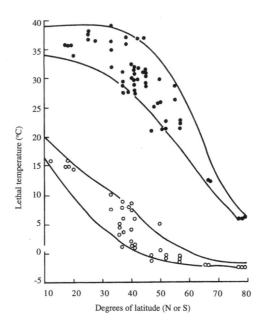

Figure 2.3. **The correlation of upper and lower lethal thermal temperature limits of fishes with latitude.** (Vernberg and Vernberg [4]).

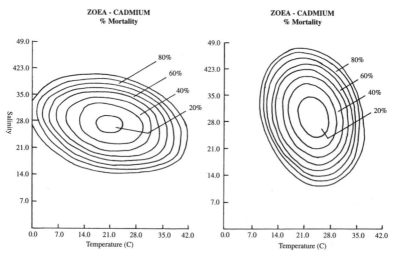

Figure 2.4. **Estimation of percentage mortality of first stage *Uca pugilator zoea*** based on response surface fitted to observed mortality under 13 combinations of salinity and temperature (a) with, and (b) without the addition of 1 ppb cadmium (Vernberg, et al. [23]).

range. An illustration of this point is when an uncommonly cold winter dramatically decreases the population size of certain shrimp species, resulting in greatly reduced harvesting of shrimp the next year. However, many times it is not easy to explain the causative factor or combination of factors that reduces the size of a population or a species. Sublethal, but stressful, levels of an environmental factor may interact synergistically with a sublethal level of a second factor resulting in the death of an organism (figure 4). Hence, as will be discussed in later chapters, regulatory standards for the level of toxicity of a given substance or environmental factor must consider not only one factor but also the effect of the sublethal exposure of another component of the environment (multiple factor interaction).

In addition to variation in temperature due to the various factors mentioned above, the temperature an organism experiences can vary in one location due to micro environments. Underneath a rock the water temperature may be 20°C lower than that at the top surface, when it is exposed to intense sunlight. Temperatures at the surface and at various depths in the soil may also vary, and micro thermal stratification may also be observed in the surface layers of bodies of water.

Temperature is significant to both biotic and abiotic factors. Many physiological processes of an organism are directly controlled or indirectly influenced by temperature. Although an organism might survive exposure to a wide range of temperature, reproduction may occur only within a narrow temperature range (table 1). The metabolic rate of many organisms is thermally influenced. Generally the rate is low at low temperatures and higher at elevated temperatures. Moreover, abiotic factors are also influenced by temperature; the solubility of dissolved gases and the viscosity of seawater vary, for example, inversely with temperature.

Salinity

Seawater consists of a solution of inorganic salts, organic substances (both natural products and manmade compounds), and dissolved atmospheric gases. The amount of salt in a sample of seawater is expressed as *salinity* and is written as parts per thousand (ppt) or o/oo. Waters in the coastal zone can vary in salinity from 0 o/oo to over 150 o/oo. The salinity of open ocean seawater, however, is about 30–35 o/oo and varies only slightly. However, the salinity of coastal waters, especially those receiving large inputs of riverine water, may be lower and variable. Within an estuary, gradients of salinity may be observed, ranging from 0 o/oo where freshwater enters the estuary to 35 o/oo at the estuary/oceanic interface. Values higher than 35 o/oo are found in shallow, intertidal pools subjected to the high temperatures and intense

TABLE 2.1

Upper limits of temperature range during breeding season in different species of barnacles (5).

Species and Geographical Range in Europe	Breeding Season	Range of Monthly Mean Sea Temperatures at Southern Limit of Range During Breeding Season (C°)	Upper Temperature Limit (C°)
B. *balanoides* Arctic–N. Spain	Nov–Dec	11–13	14–16
B. *balanus* Arctic–English Channel	Feb–Apr	9–10	13–14
V. *stroemia* Iceland–Mediterranean	Jan–Apr	13–15	21–23
B. *crenatus* Arctic–W. France	Jan–June	11–16	22–124
C. *stellatus* N. Scotland–W. Africa	Probably throughout year	19–27	29–31
B. *amphitrite* W. France–Equator	June–Aug	27–29	>32

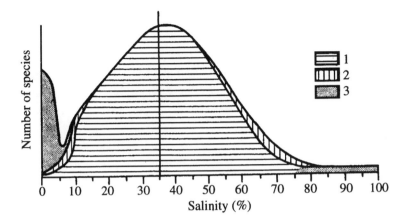

Figure 2.5. **The origin of animal species inhabiting an estuary and their distribution along a salinity gradient:** (1) number of marine species, (2) brackish water and euryhalinic estuary-living species of marine origin, and (3) freshwater-living animals or their descendants (Vernberg and Vernberg [5]).

sunlight typical of tropical salt flats. Along the coasts of the United States, especially in the South, shallow bays with poor circulation may have higher salinity values than oceanic water and are called *hypersaline* waters. In an estuary the salinity may also vary with the phase of the tidal cycle: during the incoming tide the salinity will be high because of the influx of oceanic water, while during the falling tide the value will generally be lower, reflecting freshwater input from river systems. As a result of living in a fluctuating salinity environment, long-term residents of an estuary are adapted to withstand wide variations in salinity, unlike most oceanic species that live in a more-or-less stable salinity environment (4, 5).

Salinity is important in determining the distribution of plants and animals in coastal regions. Most freshwater organisms carried by river systems to estuaries will not survive exposure to higher salinities. Many studies indicate that salinities from 5 to 8 o/oo limit seaward distribution of freshwater species, and this range limits the landward distribution of marine species up estuarine-river systems (figure 5). For example, for animals associated with an oyster-bed community, many more species will be found in high-salinity water than in low-salinity water (6). Similar results have been found for various species of fish and other groups of invertebrates and plants (5).

Much variation in the response to salinity change has been observed in estuarine organisms. For example, oysters can tolerate both low and high salinity water for considerable periods of time, a *euryhalinic* response. By contrast, the bay scallop is not very tolerant of low salinity—its tissues die at about 15 o/oo. Organisms that survive only over a limited range of salinity are called *oligohalinic*. Salinity may not only have a lethal effect on an organism, but a change in salinity may also result in a negative sublethal response. For example, reduced but sublethal salinity may inhibit the reproductive capability of a species as well as its ability to feed and metabolize. Obviously if a species is unable to reproduce or have a life-sustaining diet, the species will gradually disappear from an area. Hence, if dredging alters the salinity of a region, an immediate mass mortality of organisms (lethal effect) may not be observed, but with time a gradual diminution in the population size of the species (sublethal effect) will occur (5).

Dissolved gases

Although seawater contains thousands of different natural and "human-made" chemical substances, special attention must be given to the dissolved gases because they greatly influence many of the vital functional processes of both individual organisms and the entire ecosystem. In the absence of oxygen (*anoxia*), one of the dissolved gases, most organisms will perish. The point at which an organism dies from lack of oxygen varies with species and even differs with various stages in the life history of an organism (4, 5). For example, some polychaete worms that normally live in muddy, low-oxygen habitats can survive better in low-oxygen water than species residing in oxygen-rich waters. Not only will anoxic waters dramatically influence the survival and distribution of an organism, but when oxygen levels drop below normal values (*hypoxia*), its physiology is affected. In the above example, reduced oxygen concentration inhibited the reproductive capability of the polychaete worm but not its ability to survive. The importance of dissolved oxygen is recognized by regulatory agencies; it is one of the key water quality parameters used in regulating pollution.

The concentration of dissolved gases, such as oxygen, is influenced by a number of abiotic and biotic factors. Oxygen is introduced into coastal waters by gas diffusion across the interface between air and water and by photosynthesis. These processes are limited to the uppermost layers of the water column, but vertical mixing of the coastal waters distributes dissolved oxygen throughout the water column. Other environmental factors to be considered are temperature and salinity. High temperature and high salinity each decrease the solubility of oxygen. By contrast, low temperature and low salinity increase solubility of oxygen (at 0°C oxygen solubility is 14.6 milligrams per liter; at 35°C it is 7.1 mg/liter). Details of measurement of dissolved gases, including tables of solubility, may be consulted (27).

When increased amounts of organic material are introduced into waterways (*organic loading*), this material creates a demand for oxygen (*biological oxygen demand*, or BOD), resulting in decreased concentrations of dissolved oxygen. Further, as temperature increases, organisms typically increase their metabolic rates, further decreasing the amount of available dissolved oxygen. This competition for oxygen can have adverse effects on coastal aquatic ecosystems by creating anoxic or hypoxic conditions (see chapter 7 for examples of anoxic environments). The solubility of other dissolved gases also influences the ecological dynamics of the coastal biota (as when decreased nitrogen solubility contributes to an alteration of primary production (9, 10, 28).

Tides

One of the most distinctive characteristics of the coastal zone is the presence of an intertidal zone that experiences rhythmic fluctuations in sea level known as *tides*. The type of tide varies in different parts of the world: when high water occurs approximately once a day (averaging 24.8 hours), the pattern of tidal change is called a *diurnal tide*, but when high water occurs twice daily at intervals averaging 12.4 hours, it is a *semi-diurnal tide*. The magnitude of tidal change varies geographically, ranging from a few inches in some areas to more than forty feet in such places as the upper reaches of the Bay of Fundy. Also, tidal fluctuation varies on a regular time basis. At the times of the new and full moon when the sun, moon and earth are directly in line, the magnitude of tidal change is maximum, causing very high and very low tide levels. These phenomena are called *spring tides* (spring tides are not to be associated with the spring season but occur about twice a month throughout the year). When the moon, earth, and sun are at right angles to one another, the magnitude of tidal change per tidal cycle is minimal and is known as a *neap tide*. The gravitational forces of the moon and sun are additive during spring tides, but they tend to cancel out each other during neap tides. The tidal range varies with other factors as well, such as wind velocity and direction. For example, if a strong wind is blowing in the opposite direction of the incoming tide, the height of the high tide will be lower than predicted.

In parts of the world where the tidal changes are great, the intertidal zone has been subdivided into various subunits (7). Mean sea level (MSL) is the average level of the sea above a reference point called a chart datum, which is arbitrarily selected so that few tides fall below it. Mean tide level (MTL) is the average height of mean high and low waters determined over a long period of time. Mean high water spring (MHWS) is the average height of high water during spring tides. Mean high water neap (MHWN) is the average high water level during neap tides. Mean values at spring and neap tides have also been determined. Certain species are restricted to specific regions of the intertidal zone, but others may extend over the entire intertidal zone depending on their ability to withstand the environmental complex typical of the different intertidal zone regions. For example, organisms living in the highest intertidal zone are exposed to air temperatures for a longer time than organisms residing at lower levels. Also organisms in the higher intertidal zones are likewise subjected to longer periods of sunlight and shorter periods of coverage by seawater. Obvious physiological adaptations between species limited to these different zones exist, and any human-induced alterations in tidal levels of estuaries that impair the organism's physiology can profoundly influence the distribution of intertidal plants and animals.

Light

Light is a very dynamic abiotic factor whose changes may be both rhythmic (daily and seasonal) and arrhythmic (transitory changes due to differences in cloud cover). Without light, photosynthetic activity would not be possible; fluctuations in the amount of light can influence the primary production level. In addition, light directly influences animal behavior and physiology. Various aspects of light, including its intensity, wavelength, polarization, angle of reception, and duration (*photoperiod*), have marked ecological effects. Although light is important to land—dwelling organisms, it also exerts a strong influence on estuarine species. For example, the amount of sediment (*particulate*) material in estuarine waters influences the amount of photosynthesis—increased sediment loading decreases photosynthetic output as light penetration through the water column is decreased.

Substratum

The types of sediment or bottom substrata found in the coastal regions are extensive and varied. The physical and chemical properties of each type of substratum are very important to the distribution of terrestrial and aquatic species that are intimately associated with it. By contrast, aquatic organisms living in the water column or terrestrial species which are mobile are more independent of the substrata. Certain species are found only on or in one type of substratum, and if that type is altered either by natural and (human) anthropogenic processes, that species will be displaced. For example, the ditching and draining of coastal wetlands may drastically alter the soil type, resulting in the disappearance of existing wildlife and trees and vegetation. In general, probably because of the relative economic importance of uplands for human uses, more detailed maps of the various types of coastal zone soils are available than are those for the intertidal and the coastal subtidal zones.

Two principal types of substrata are recognized: solid and particulate. Solid substrata include rocks, wood, shells, and even beer cans, while particulate substrata are sand, mud, clays, and mixtures of each. Particulate material is easily transported from one location to another by air currents or flowing water, in contrast to the relatively stable behavior of solid substrata. Riverine systems carry a tremendous amount of sediment originating from upland runoff to downstream estuaries. These sediments not only accumulate in harbors, where they necessitate dredging to maintain shipping channels, but they also may carry various chemicals and microorganisms attached to them. Such chemicals may be toxic to species or, in some cases, the attached chemicals may be harmless or beneficial. Airborne particulate matter also influences the environment, causing serious health problems, and, in certain regions, alterations in soil types. This transformation may result in the change of a habitat which, in turn, influences the species composition of plant and animal communities: transport of sand particles, for instance, may cause the movement of sand dunes onto agricultural fields, making them unproductive.

Sound

Sound influences terrestrial as well as aquatic animals. Not only does sound serve to attract or repel an organism for such important physiological functions as mating, eating, and predator avoidance, but excessive exposure to sound has a negative effect, as when species are adversely affected by being too close to busy airports. Sound travels five times faster through water than through air, with its velocity affected slightly by salinity, temperature, and pressure.

Biogeochemical cycles

An essential active process within an ecosystem is the cycling of chemical substances through both the biotic and abiotic components. This process involves passage through the biotic phase (the various trophic levels) and subsequent release from an organism, either as waste products or upon death, to an abiotic phase when the chemical compounds or elements then become available to other organisms, especially primary producers. Because

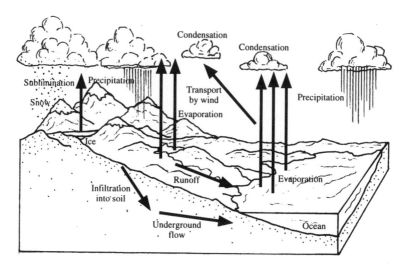

Figure 2.6. **The hydrologic cycle.** Snow and rain together equal precipitation. Sublimation, the direct transfer of ice to water vapor, is included under evaporation. Surface flow and underground flow are considered as land runoff.

this cycle involves a biotic phase and a geologic (abiotic) phase, it is termed a *biogeochemical* cycle. Although cycling of chemical substances has been recognized for years, newer technologies have been developed to permit a better quantitative and qualitative understanding of the fundamental importance of these cycles. Certain cycles have been known and studied longer than others. For example, the carbon, nitrogen, oxygen, and phosphorous cycles were known earlier than were cycles of other substances such as chemical pollutants. Because all substances must be in solution to be physiologically useful, water is essential for the function of various cycles as well as being important in itself as a cycling compound. In general, not all chemical elements cycle at the same rate, nor is the rate the same for one element in various environments—a factor that obviously increases the difficulty in understanding the implications of various cycles in the application of management practices.

To describe a cyclic phenomenon, it is necessary to enter the cycle at one point and follow the process completely around to that same starting point. Some of the better known cycles are presented below.

Hydrologic cycle

Figure 6 graphically depicts the transfer rates of water in various forms between different regions of the planet Earth. The global transfer of water

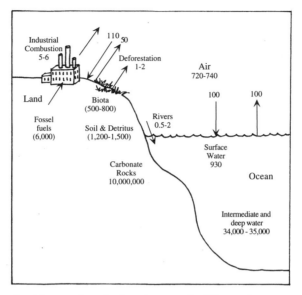

Figure 2.7. **The biogeochemical carbon cycle.** The carbon content, measured in billion metric tons (bmt), and fluxes (bmt/year) in and among the earth's major reservoirs are estimated in this model. The natural carbon exchange between the fossil-fuel reservoir and the atmosphere is slow compared to the rate of interchange between the biosphere and the atmosphere. (modified from Pinet [24]).

has great significance to society in that a shift in transport of water from one geographical region to another determines, in large part, the productivity of agricultural crops, resulting in drought or excessive rain. Not only is it necessary that human society develop a better grasp of the hydrologic cycle on a global basis, but it is imperative also to have an accurate understanding of the dynamics of water fluxes on a regional and local basis. Many coastal areas are experiencing shortages of water as a result of increased development. For example, developmental practices which increase surface runoff decrease infiltration in the soil and can reduce the size of the groundwater pool. In addition, with development water use occurs at a faster rate than it is being replaced in underground reserves. Water is not an unlimited resource, and it is basic to survival of coastal ecosystems.

Carbon cycle

One of the most prominent environmental problems facing human society concerns the global cycling of carbon, especially the role of carbon dioxide

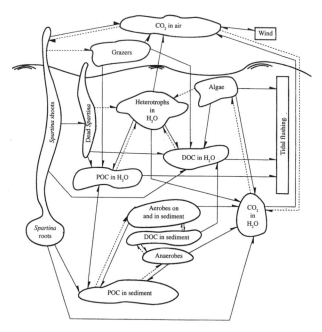

Figure 2.8. **Fourteen-component carbon-flow model of a coastal Spartina alterniflora salt marsh** (modified from Wiegert and Freeman [25]). POC=particulate organic carbon; DOC=dissolved organic carbon; CO_2=carbon dioxide.

(CO_2) in determining global temperature (figure 7). Data indicate a steady increase of atmospheric CO_2 in recent years due to increased use of fossil fuels, especially in developed countries, and deforestation on a worldwide basis. Burning of fossil fuel results in a number of by-products including CO_2, whereas deforestation influences the carbon cycle by halting the removal of CO_2 by trees in the photosynthetic process. Based on various models, this shift in part of the carbon cycle appears to have resulted in an increase in mean global temperature. The implications of global warming are discussed in chapter 7.

Another important part of the carbon cycle involves plants and animals. Carbon is a major constituent of living organisms. Basically plants convert CO_2 into carbon-containing molecules during photosynthesis. These organic molecules are incorporated into plant tissues which may then be passed to higher trophic levels via food-web dynamics. In turn, organic carbon is converted to CO_2 as a product of respiration by microorganisms during decomposition of dead matter. Figure 8 depicts the dynamics of carbon

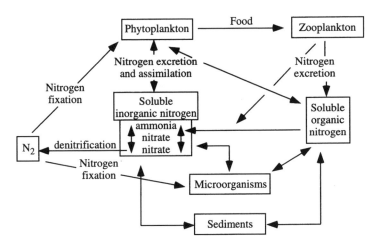

Figure 2.9. **The nitrogen cycle in the sea** (From Meadows and Campbell [26]).

cycling in an estuary. In the coastal zone, carbon compounds can be transported from one habitat to another—as when carbon is dynamically exchanged between estuaries and the coastal waters as a result of tidal action (8). Some carbon is incorporated into inorganic carbonate in shells both of larger organisms (such as snails, clams, and oysters) and smaller microorganisms that settle to the bottom of oceans to form biogenic ooze. Another important source of carbonates results when corals secrete massive coral reefs. Unlike the metabolic role of carbon, which occurs over short periods of time, the turnover of carbonates in the carbon cycle occurs on a geologic time scale.

In addition to the fundamental role of biological activity in the carbon cycle, the ocean serves as a large reservoir with a large storage capacity. Carbon exchange during the complex interaction between the atmosphere and water environments is poorly understood and is under active study.

Nitrogen cycle

The various pathways of cycling nitrogen on global and regional scales, involving various biotic and abiotic components of an ecosystem (figure 9), are complex and not fully understood. However, it is known that wetlands play an extremely important role. Increased anthropogenic input of nitrogen oxides and ammonia to the atmosphere increases nitrogen availability to plants, with the potential effect of influencing the productivity and structure of coastal ecosystems. Although exceptions can be found, in general, primary production in wetlands is limited by nitrogen (9). Increased additions to wet-lands via atmospheric nitrogen and surface runoff of fertilizers from uplands can selectively stimulate production by nitrogen-loving plants, which in turn can favor these plants in competing with other, less nitrogen-dependent plants. This shift in the structure of the wetland community, as a result of competitive advantage, results in decrease or extinction of certain plant species. In turn, the population composition of species that are dependent on the less competitive plant species will be altered. Additional information on nitrogen cycling can be found in a number of recent articles (9, 10).

Other cycles

Other cycles exist —for instance, sulfur, phosphorus, calcium— are of eco-logical importance to the coastal environment. The reader is urged to con-sult other sources for detailed discussions (11).

Ecosystem structure and function

The structural anatomy of a system is based on knowledge of both its biotic and abiotic components. Descriptions of the biota include such basic data as number of species and the number and size of individuals (if expressed as weight, it is called *biomass*), seasonal changes in appearance or number, and the distribution of these in time and space.

Quantitative and qualitative chemical analyses of both living and non-living substances are necessary. Such physical parameters as soil types; rates of sedimentation; temperature; rainfall; solar radiation; salinity; and circula-tion patterns of streams, estuaries, and coastal water are important to analyze on a temporal and spatial basis. In addition to knowing the structural com-position of the system, it is also important to understand the function and dynamics of the workings of the system. A few examples will illustrate this point. How do abiotic and biotic factors influence the photosynthetic process? What is the role of light and sediment concentration on organic production? How fast are nutrients recycled? Does an increase in salinity influence the functioning of the various species in the same way? What is the rate of energy transfer from one trophic level to another under fluctuating environmental conditions?

Knowledge of how ecosystems function is of fundamental value to the scientific subdiscipline of systems ecology and also has practical applications to the field of environmental management. One of the most intriguing sci-entific questions today is how an ecosystem functions and changes over time. What are the impacts of human interventions on the ecosystem? How do we manage the environment wisely to prevent irreversible damage and a resulting negative impact on the quality of human life? What changes are needed to restore damaged habitats to their productive, pre-impact status? What types of ecosystem models can be developed to predict the short-term and the long-term impacts of environmental change? Citizens involved in the decision-making process must have knowledge of the potential effects of altering the environment on ecological systems. One of the goals of systems science is to provide a knowledge base for rational decision-making rather than relying on emotions and special interests.

Ecosystem productivity

One feature of the coastal zone that attracts human settlement is its high level of biological productivity. An abundant supply of seafood, fertile land for agricultural and forestry activities, and shells are just a few examples of products of the coastal zone that are valued by our society. Understanding how the ecosystem can reach and maintain a high and sustainable level of biological productivity is essential to managing coastal resources. For purposes of presentation, ecological productivity will be divided into two categories: *primary production* and *secondary production*.

Primary production

Different plant species contribute to the total photosynthetic production in various coastal zone habitats. The terrestrial upland plant communities bordering coastal wetlands exhibit extreme diversity, varying from boreal coniferous forests to tundras to farmland to tropical mangroves. Within the nonmarine coastal habitats, we find freshwater swamps, lakes, and streams inhabited by various aquatic species. Estuaries also have several groups of plants that contribute to primary productivity. Along the eastern coast of the United States, the vascular plant *Spartina alterniflora* is a dominant biotic component covering most of the estuarine intertidal zone. By contrast, in European estuaries vascular plants tend to be limited to the upper reaches of the intertidal zone, with eelgrass and algae being dominant in the lower intertidal zone. Other types of primary producers are macroalgae (seaweeds), sediment-dwelling diatoms, and phytoplankton.

In addition to the great biological diversity exhibited by coastal zone plant species, the amount of organic carbon produced by these species in the major habitat types varies. Table 2 demonstrates the general differences in production estimated for different ecosystems (12, 13). Deserts and tundras have the lowest values, with estuaries and wet tropical and subtropical forests being most productive on a unit per area basis. Within an estuary, the annual primary production contributed by the various types of plants is different

TABLE 2.2

Net primary production in various major habitats (from various sources).

Habitat Type	Area (million km²)	Net Primary Productivity per unit area (g/m²/yr)*	
		Range	Mean
Open Ocean	332.0	2–400	125
Estuaries	1.4	200–3500	1500
Lake and Stream	2.0	100–1500	250
Tropical Rain Forest	17.0	1000–3500	2200
Temperate Deciduous Forest	7.0	600–2500	1200
Boreal Forest	12.0	400–2000	800
Temperate Grassland	9.0	200–500	600
Cultivated Land	14.0	100–3500	650
Desert and Semidesert Scrub	18.0	100–250	90

*grams per square meter per year

TABLE 2.3

Annual net primary production of the North Inlet estuarine system, South Carolina (14).

	Annual net production (gCm²yr)*	Percentage of net production
Spartina plant community	400	40
Macroalgae (seaweeds)	200	20
Microphytobenthos (Mud diatoms)	200	20
Phytoplankton	100	10
Epiphytes and neuston	100	10
TOTALS	1000	100

*grams of carbon per square meter per year

with phytoplankton, epiphytes, and *neuston* (organisms inhabiting the water surface) being less productive than *Spartina* (see table 3). The same habitat type (that is, estuary) from different geographical regions may not have the same level of primary production (see table 4) (14, 15). Nor does one specific habitat have a constant annual production value with lowest to highest values for *Spartina* occurring during winter months (16). Even within one estuary the level of productivity by one species may vary depending on its

TABLE 2.4

Net aboveground primary productivity of *Spartina alterniflora* from different geographic areas (15).

State	Net Primary Productivity (g/m²/yr)*
Louisiana	750–2600
Mississippi	1473
Florida	130–1281
Georgia	985–2883
South Carolina	724–2188
North Carolina	329–1300

*grams per square meter per year

location. For example, a population of *Spartina* growing along creek sides has a higher amount of primary production than a population of the same species living in a midmarsh site (17). Although intersite and temporal differences in production are noted, it is important to know the relative importance of these differences when making management decisions, especially concerning proposed environmental modifications capable of influencing primary production. If less food is produced by these plants, higher trophic levels will not be able to flourish. To help explain differences in primary production, it is necessary to have an understanding of those factors which influence this important ecosystem process.

One of the key factors controlling photosynthesis is light. The amount of available light varies seasonally in many of the coastal environments with the greatest fluctuation occurring in polar regions and the least in tropical regions. Plants living in the water column are influenced by seasonal variations in light; thus the clarity of the water is important. For example, if the water contains much sediment, light cannot penetrate very deeply and the total photosynthetic output is restricted. The depth at which energy gained by photosynthesis is equal to the utilization of energy by the plants for respiration is called the *compensation depth*. If plants stay below this point for too long, they cannot survive and they literally starve from lack of sunlight. Within estuaries subjected to excessive groundwater runoff, the compensation point is near the surface of the water column, whereas in clearer water, this point is deeper. Thus the potential for a higher level of primary production is generally higher in clearer water. One of the hazards of dredging in estuarine and coastal waters is the increase in the level of suspended sediments. The resulting sedimentation decreases the clarity of the water and, in turn, the rate and amount of photosynthesis.

Numerous other factors influence photosynthesis by coastal vegetation. Although a detailed discussion of these factors is beyond the scope of this book, for the purpose of understanding how human intervention may alter the productivity of coastal regions, a few of these factors will be highlighted. In general, a reduction in temperature reduces the biological activity of plants, while up to a point increased temperature will enhance photosynthesis. Beyond these critical low or high temperature points, the plant will die. The discharge of heated effluent waters may raise the temperature of coastal waters to or beyond this thermal lethal point. In addition, inhibiting the flow of water to estuaries by upstream engineering projects (construction of dams, reservoirs, or diversion of river flow) or creating impoundments within an estuary may result in increasing water temperatures to a critical point. Not all species of plants tolerate the same range of temperature change; cold-water species, for instance, cease to photosynthesize at lower temperatures than warm-water forms. Hence, regulatory standards based on the response of plants from colder regions (such as Maine), may not be applicable to tropical plants (such as those from Florida).

High production levels in estuaries are associated with high levels of nutrients, especially nitrogen and phosphorous; length of the growing season (a function of light and temperature); increased tidal range; and sedimentation processes. Of these factors, much attention has been given to the role of increasing levels of nutrients in aquatic system. Levels of nutrients high enough to alter system dynamics by stimulating plant biomass production and reducing the level of dissolved oxygen results in a process known as *eutrophication*. Various studies have clearly demonstrated that the addition of nitrogen to marshes increases primary production many-fold. Although extensive research has demonstrated that nitrogen is added to waters from adjacent highlands, recent studies estimate that another important input of nitrogen into marshes and coastal waters is from atmospheric deposition (18). This new finding has importance in developing management plans to control nutrient input.

Secondary production

Although primary production is important to the functioning of coastal ecosystems, salt marshes are also very dependent on the role of abundant, bacterially rich detritus as a food source for many of the animals in this habitat. There is increased survival value to the estuarine animal community in having multiple sources of food, since detritus tends to be available throughout the year while primary production varies seasonally. The diversity in food availability and abundance helps explain why marshes provide an excellent habitat for the many oceanic species that invade marshes on a temporary basis as well as serving as an ideal habitat for resident animals.

Secondary producers can be divided into at least two types: those that consume primary producers directly (primary consumers or herbivores) or those (secondary consumers or carnivores) who feed on primary consumers. In general, secondary producers attract the most attention of the general public because they are conspicuous, with readily observed recreational, aesthetic, and/or commercial value. Examples include birds, shrimp, clams, and oysters. The secondary producers digest the energy-containing molecules found in the cells of the primary producer and derive energy both to sustain their life processes and to provide biochemical building blocks necessary to synthesize the proteins, lipids, and other essential molecules characteristic of their bodies.

Within the coastal zone, very complex and diverse ways have evolved for secondary producers to obtain food. The information found in the scientific and general literature dealing with this subject is too voluminous to review in this book, but some general comments are in order to understand the important role of these animals in the functioning of ecosystems (figure 2). Although more information is presented in a later chapter, it should be borne in mind that any proposed human disturbance of the coastal zone environment should be examined for its potential disruption on the feeding capabilities and the population sizes of the secondary producers.

Herbivores have several methods of feeding on primary producers. One of the most common is by filtering from the water column such important primary producers as phytoplankton. Oysters and clams are excellent examples of filter feeders. These animals remove phytoplankton and other materials suspended from the water, then process these materials and transform them into molecules that are released back into the water. These molecules are utilized by other organisms (19). In some estuaries with abundant oyster reefs, approximately 68 percent of the water can be filtered over a tidal cycle. In addition to the easily identified primary consumers, many species of small inconspicuous animals, such as zooplankton and benthic invertebrates, are filter feeders and influence ecosystem dynamics. Other herbivores do not feed on primary producers found in the water but subsist on larger vascular plants. Obviously in the upland coastal regions, cattle and other livestock feed on grasses, but these large animals are not usually found in saltwater wetlands. Here the dominant vegetation is various species of cordgrass, especially *Spartina alterniflora*. More than 600 species of insects have been described that live in this habitat, and many of these feed directly on *Spartina* (20). In addition to being eaten by primary consumers, dead *Spartina* decomposes and serves as a substratum for microorganisms and breaks up into detritus that can serve as food for consumers.

Conspicuous in coastal regions are the numerous and various types of secondary consumers. Some are migratory coastal species, utilizing coastal resources for temporary periods of time, while some are permanent inhabitants of one type of coastal habitat. As a result of this complex mix and occurrence of various species, numerous types of food webs exist. For example, about 90 percent of the fishery landings on the Southeast and Gulf coasts involve species with different food requirements that spend some portion of their lives in an estuary (21). Thus the food resources associated with estuaries are consumed not only by resident species, but by various life history stages of migratory species. Natural or human alteration of factors influencing the food resources will have an effect on the secondary consumers. Temperature, salinity, oxygen concentration, water currents, tidal flooding, and light are just a few examples of important abiotic factors that influence feeding. A number of biotic factors (competition between individuals, body size, reproductive state, and physiological needs) are equally important in determining utilization of food resources.

An important ecological generalization to bear in mind when considering how coastal systems function is that every available food source within a given habitat is potentially consumed by a consumer species or species in that locale. Studies on estuaries demonstrate this generalization: some species feed at night and some during the day, while others are continuous feeders. Other species feed on resources found in the water column; some are bottom feeders; others are parasitic.

Secondary production in the coastal zone is a complex process resulting in the production of organisms that are valued by humans for commercial, ecological, and esthetic reasons. Because of the complexity of the coastal zone, proposed alterations of coastal habitats should be carefully evaluated so that the dynamics of interrelated food webs are not changed, for change often results in unforeseen negative impacts on the system.

The concept of an ecosystem is a useful method of organizing our thoughts about how the coastal zone environment is structured and how it functions. Knowledge of the interdependence of biological and physical components of the environment is important when considering the potential impacts of human disturbance on the ecology of this vital segment of planet Earth. This coastal ecosystem is subjected to short-term and long-term natural and human-induced environmental changes that can alter trophic dynamics of the coastal biota.

3

Major Coastal Zone Habitats

The ocean environment

The earth is often referred to as the "watery planet." This feature makes it unique among planets of our solar system. About seventy percent of the earth's surface is covered by fluid water, and the quantity of seawater is so great that if the earth's surface were flattened, it would be covered with water to a depth of 7,350 feet. In addition to being a distinct ecological subdivision of earth, the sea exerts a strategic influence on the ecology of the total biosphere. It is a well established fact that climate is profoundly influenced by the sea. Although little is known of the actual cause-and-effect relationships, the physical properties of seawater allow it to moderate the climate: fluctuations in oceanic currents can alter the weather of adjacent land masses, as demonstrated by the El Niño phenomenon. It is also important to note the role of the sea in the hydrologic cycle, providing an important source of water essential to all forms of life.

The marine environment is not homogeneous but varies geologically, physically, chemically, and biologically. The greatest diversity of major animal types exists in the ocean, an indication of its fitness to support biotic activity. Furthermore, all life apparently originated in the sea, and the amount of living matter in the ocean far exceeds that on land. Unlike terrestrial or freshwater systems, which may be isolated from one another, all the great oceans of the world are interconnected and flow continuously in a state of dynamic interchange. Theoretically, a small drop of seawater from a specific site in the coastal zone could, in time, reach any other part of the marine environment. As noted earlier, although the salinity of open oceanic water is about 35 o/oo, this value may vary greatly in waters of the coastal zone, ranging from 0 to more than 90 o/oo. Temperature fluctuations in the ocean over time are less extreme than on land or in coastal waters. Because it covers so much of the surface of the earth, the ocean receives most of the radiation from the sun and thus serves as an immense reservoir of heat. In turn, water vapor is given off to the atmosphere, and the ocean thus contributes to the reservoir of heat in the atmosphere. To facilitate the study of the marine environment, classification of the various marine habitats has been proposed (figure 1)(1). Based on a number of criteria, two major divisions, *pelagic*

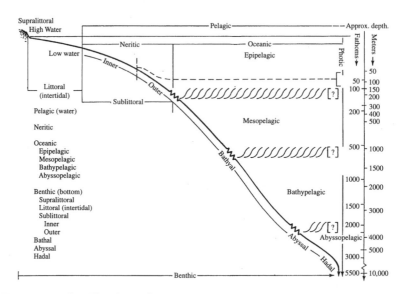

Figure 3.1. **Classification of marine environments** (from Hedgepeth [1]).

(water) and *benthic* (bottom), have been identified. As depicted in figure 1, each is divided into a number of subdivisions that vary with depth and other factors such as temperature and light.

Intertidal zone

Beginning from the shore and progressing seaward, a number of distinctive habitat types may be observed. One of these types is the intertidal zone, which encompasses the area between the high-tide mark and the low-tide mark. This habitat has received much attention because it is readily accessible for humans to reach and observe and it does not require elaborate oceanographic equipment to study. However, by contrast to the more stable deep-sea habitat, it is an extremely complex environment to analyze because it is exposed alternately to aquatic and aerial factors that fluctuate on a daily and seasonal basis. Although tidal range depends on such factors as phase of the moon, the amount of tidal change varies greatly. For example, in Canada's Bay of Fundy the change may be as great as 15.4 meters, whereas in the Caribbean the change may be measured only in centimeters.

The substratum, which is very important to animals and plants as sites for attachment and shelter, forms the basis for the classification of intertidal-zone habitats. Thus we speak of sand beaches, rocky shores, or mud flats. Within each of these intertidal zone habitats there is conspicuous zonation of organisms relegating them to different parts of the intertidal zone from the high-tide mark to the low. Some animals, such as the ghost crab, will be found primarily above or near the high-tide mark, while other species, such as the box crab, are found only near or below the low-tide mark. These differences in zonation reflect the physiological and competitive capacity of organisms to respond to both abiotic and biotic factors in the environment.

Sand beaches

One of the best known intertidal areas is the sand beach (plate 1, following p. 48). These beaches have fairly uniform topographical gradations with few sharp irregularities. Two general types have been described: the open beach, which is unprotected from unrelenting wind and wave action; and sand beaches found in harbors and estuaries, which generally are protected from the pounding action of the surf. Unprotected beaches particularly are in a constant state of flux, and they may vary in extent from tide to tide, season to season, and storm to storm. Organisms inhabiting this zone must be able to adapt to a variable and harsh environment, for they are subject to desiccation, insolation, and wide temperature ranges. The size of the sand grains, which has a marked influence on the physical characteristics of the beach, may vary not only from beach to beach, but also at different regions of one beach. This variation in sand grain size plays an important role in the distribution of animals. Some species prefer fine sand grains, while other species inhabit coarse sand habitats. Temperature and salinity at the surface of sand beaches fluctuate more than that at depth.

Rocky shores

Rocky shores also characterize the intertidal zone habitat (plate 2, following p. 48). Rocky intertidal habitats, where the substratum is a hard, solid surface, vary greatly. Some are uniformly flat from the sea to land, whereas others are characterized by boulders of varying size with irregularly shaped surfaces. In this latter type many microenvironments exist, accompanied by a rich variety of fauna and flora. Owing to marked differences in various environmental factors, organisms on rocky shores exhibit a high degree of zonation. Associated with rocky shores are tide pools and the splash zone, the latter being a region above the high-tide mark that is splashed with seawater as a result of considerable wave action. Organisms living on rocky shores must contend with problems of both an aquatic environment when the tide is high and an aerial environment at low tide. Rocky-shore organisms generally have some mechanism to attach firmly to the surface of the rocks or the ability to grasp the substratum so that they are not washed away by the powerful action of the smashing waves.

Mud flats

Mud flats represent another dominant intertidal zone area (plate 3, following p. 48). Here there is a gradation in habitats ranging from pure sand to pure mud, depending upon the amount of sand and the size of the sand particles. Typically, mud cannot exist where there is significant wave action; hence this type of habitat generally occurs in protected waters such as those found in estuaries. Since mud is of fine texture and has high colloidal content, there is little interstitial water, or water exchange, and the oxygen content is low, but there is likely to be a high organic content that can serve as a rich food supply for organisms living in it. Usually a black layer of mud exists a short distance below the surface, indicating low oxygen and a high hydrogen sulfide content that gives off the odor of rotten eggs. Since the substratum is soft, most organisms living there are burrowing animals. Except in those areas of a mud flat where shell or other hard materials are mixed in, few attached organisms are present.

Adjacent to sand and mud flats in areas influenced by salt water, extensive salt marshes may be found. Salt marshes are replaced by freshwater wetlands further up riverine systems. Because salt marshes and freshwater wetlands are associated with estuaries, an ecological characterization of these habitats is presented in this chapter under the section on estuaries.

Other intertidal zone habitats

Other habitats found in the intertidal zone include pilings of piers, seawalls, or other structures associated with human activities. Organisms may attach themselves to these structures. And if the structures are constructed of wood, a whole host of wood-burrowing organisms may invade them and cause extensive damage. Irrespective of the type of intertidal zone, debris accumulates above the high-tide mark when the tide recedes. Although the size, shape and composition of this debris varies, a relatively complex community of organisms lives here in what is appropriately called the *drift-line habitat.*

Estuaries

Wherever water from a freshwater system pours into the sea, a physically variable environment is created. This region of dynamic encounter is referred to as an estuary (plate 4, following p. 48), "a semi-enclosed coastal body of water which has a free connection with the open sea and within which seawater is measurably diluted with freshwater derived from land drainage" (2). Although other definitions have been proposed, all emphasize that estuaries are restricted to the mouths of rivers in areas where oceanic tidal fluctuation occurs.

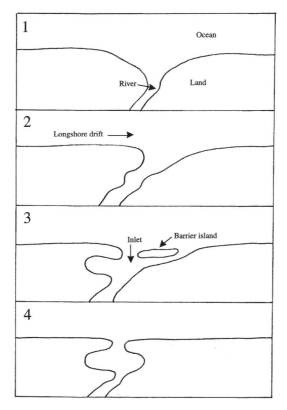

Figure 3.2. **Successional stages in formation of bar-built estuary.** (1) A river enters a coastal area at the ocean. (2) The longshore drift current deposits sand on land at the mouth of the river. (3) Sand is deposited to form a barrier island. Additional sand is deposited on the opposite shore forming embayments. (4) The barrier island becomes attached to mainland, forming a well-defined opening between ocean and the embayments.

Various types of estuaries are found along the coast of the United States. In the northern West Coast and in New England, fjords occur. A distinctive feature of this type of estuary is its formation by "gouging" glacial action. The presence of a sill across the estuary's mouth, which inhibits the circulation of water found in the lower depths of the fjords, results in such events as stagnation of water and concentration of pollutants. Another type of estuary found on the West Coast results from slippage of tectonic plates, as in San Francisco Bay. A third type of estuary results when the sea level rises and floods river valleys (Chesapeake Bay, for example). Numerous examples of the fourth type of estuary, the bar-built estuary, can be found in southeastern

United States coastal regions. Bar-built estuaries result where barrier islands are formed from near shore sand bars (figure 2). With time, channels cut across or around barrier islands, allowing tidal interchange between oceans and riverine water.

The upper limit of an estuary is difficult to pinpoint. In some estuaries, such as the Amazon River, tidal influences are observed several hundred miles upstream, well beyond the presence of saline water. In general, estuaries experience a change in salinity associated with the tidal cycle. However, some estuaries experience relatively little change, depending on freshwater inflow and the pattern of water circulation within the estuary. The principal types of estuarine circulation are presented in figure 3.

Salinity has a profound influence on the distribution of many animals. As depicted in chapter 2 (figure 2), the number of marine animals declines in direct proportion to decreasing salinity, with the greatest number of species found near shore in undiluted seawater. Alterations in salinity due to human interventions (such as damming of rivers) can dramatically influence the kinds of organisms inhabiting estuaries.

In addition to species that spend most of their lives in an estuary, many oceanic species utilize the estuary only for an essential phase of their life cycle. Of particular economic importance to humans, most commercially important species are dependent upon the estuary, and decreased environmental quality of estuaries results in decreased population sizes of these tasty species.

Riverine waters flowing into estuaries carry vast amounts of silt, much of which is deposited on the bottom, resulting in harbors filling in and becoming too shallow for ship traffic. To remedy this problem, extensive and recurring dredging operations are required (see chapter 7 for a detailed discussion of problems associated with dredging). In addition to silt, many other substances are carried by rivers into the coastal area. Many of these substances (such as chemicals and nutrients) result from upland runoff or direct input into rivers by human activities. Often such activities are located many miles inland from the coastal region; however, the composition of the runoff can determine which economically important species survive in an estuary. Estuaries not only have extreme economic importance in themselves, but the numerous distinctive types of habitats in an estuary and along their borders play a strategic role in sustaining the entire coastal ecosystem.

Coastal wetlands

One of the predominant coastal habitats in the United States is the marshes bordering estuaries and riverine systems (plate 5, following p. 48). The

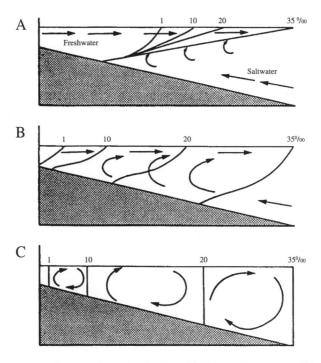

Figure 3.3. **Types of estuarine circulation.** (a) Salt wedge estuary. River flow to ocean dominates, resulting in freshwater being layered over heavier sea water on bottom. The salinity, ranging from 1 to 35 0/00, is marked off in lines called *isohalines*. Each lines represents one salinity, and the position of that salinity in the water column is indicated. (b) A partially mixed estuary. River flow is modified by tidal currents, resulting in greater mixing of fresh and salt water. (c) A vertically homogeneous estuary. Tidal currents predominate resulting in a well mixed water body. (From various sources.)

coastal intertidal emergent wetlands, located between aquatic and terrestrial habitats, comprise approximately 4.4 million acres. It is estimated that southeastern tidal wetlands represent 78 percent of the total coastal marshes in the United States (3).

Along temperate zone estuaries, wetlands extend from the sublittoral zone to the spring high-tide mark. In the tropics, salt marshes are typically confined to salt flats landward of mangroves (4). In general, marshes have several characteristics in common.

- Seaward of the marsh exists a bare mud or sand flat and landward is an area of low-growing vegetation that abuts upland vegetation (figure 4).

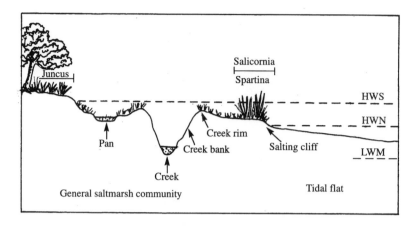

Figure 3.4. **Generalized transect of a salt marsh.** The highlands are dominated by various plant species that are sensitive to salinity. Frequently a stand of *juncus* (a plant tolerant of lower interstitial salinity) is located along the edge of the marsh where it is irregularly exposed to estuarine water. Seaward is a plant community consisting of various species living in an area between the high-water spring (HWS) levels and the high-water neap (HWN) tide mark. *Spartina*, especially *S. alterniflora*, dominates the salt marsh. Tall growth of *Spartina* occurs on edge of creek banks and on levees; intermediate size plants are found beyond the banks, and short grass is seen at higher elevation, where duration of inundation is several hours. Tidal flat is area exposed when tide recedes. LWM is median low water point. Mud flats typically are devoid of observable vegetation. By contrast, tidal salt flats may have succulents (such as *Salicornia*) and/or salt grass (such as *Distichlis*).

- Characteristically, only a few plant genera dominate North American salt marshes (*Spartina, Juncus, Salicornia,* and *Plantago*). Extending up the salinity gradient from the ocean to rivers, however, salt-marsh vegetation changes and the transition to freshwater swamps is characterized by macrophytes of the genera *Typha, Scirpus,* and *Phragmites* (4).
- Numerous drainage creeks and subunits of river systems bisect salt marshes.
- Survival and growth of marshes are dependent on the availability of silt, and protection from erosion by high-energy waves is necessary to permit sedimentation of silt.

Sediment accretion must occur at a rate compatible with colonization by salt marsh vegetation for salt marsh growth to occur. If sedimentation is too

1. Sand Beaches

2. Rocky Shoreline

3. Intertidal Mudflat with Oyster Reef

4. Estuary

5. Salt Marsh

6. Mangrove Swamp

7. Coral Reef

rapid, the area may emerge from the intertidal area and be flooded less frequently, receiving less sediment. By contrast, regions experiencing subsidence or sea-level rise may be flooded more frequently and receive more sediment. An example of the imbalance between sedimentation rate and marsh growth is seen in parts of the marshlands along the Gulf Coast of the United States where subsidence is resulting in marsh erosion.

In geologic time, a given estuary exists for a relatively short period, from thousands to a few tens of thousands of years. Various factors acting over time, including sea-level changes, subsidence, and sedimentation changes, contribute to the life and death of estuarine systems. In addition to these longer term geological changes, human perturbations can have profound negative effects on estuaries within a rather short time frame (days to years).

As indicated, sedimentation rates must equal or exceed rates of sea-level rise for salt marshes to persist. Estimates for various East Coast marshes indicate active accretion rates between 1.5 mm per year and 51. 8 mm per year, exceeding local sea-level rise in some estuaries (5–8). The annual accretion of coastal marshes must be accounted for by the net sediment transport per tidal cycle between the vegetated marsh and the adjacent tidal creek, integrated over a year cycle.

To understand how salt marshes function, knowledge of the complex interaction of physical factors with the chemical and geological components of this ecosystem is necessary. An important physical factor is the rhythmic water movement over the marsh surface, alternating between periods of flood tide followed by exposed periods during ebb tide. This water movement will influence sedimentation, chemical transport, temperature, salinity, light, and distribution of the biota, to cite a few examples. In turn, water circulation is affected by freshwater input, wind, tidal action, and geomorphology.

As discussed earlier, tidal amplitude varies temporally between neap and spring tides, with the greatest range occurring between spring low tide and spring high tide. In addition, increased wind velocity can result in greater or prolonged tidal changes depending on wind direction. In addition to these predictable changes associated with the tidal cycle, anomalous water levels can occur on time scales of days to weeks to seasons. The biotic impact of these anomalies in mean sea level (MSL) can influence ecosystem dynamics. In one study, if the transitory elevation in MSL occurs during the growing season, the annual aboveground productivity of *Spartina* was observed to have increased by a factor of two (9). In turn, the commercial landings of two estuarine-dependent species (shrimp and menhaden) were correlated with sea-level anomalies. An increased MSL may cause more of the marsh to be covered by tidal waters and lower pore-water salinity in the higher elevations of the marsh because less evaporation occurs. The decrease in salinity stimulates primary production of *Spartina* and the increased amount of submerged marsh vegetation provides more habitat area for organisms dependent on marsh grass for food and predator avoidance.

The salinity of water covering salt marshes varies depending on a number of physical factors, including rainfall; freshwater inputs from surface runoff from adjacent uplands, groundwater influx, and rivers and streams; extent and frequency of tidal flooding; evapotranspiration; soil type; and vegetation. If the salinity remains below 5 o/oo, the salt marsh vegetation is typically replaced by freshwater species. High temperatures can increase evaporation of water from shallow pools and upper portions of the intertidal zone, resulting in salinities higher than that of seawater. Temperature not only directly affects the physiological responses of the biota, but also interacts with other abiotic factors. For example, the thermal regime differentially influences the density of seawater, the concentration of dissolved oxygen, hydraulic conductivity, and the rate of chemical reactions.

Salt marshes and the overlying water column represent a complex abiotic environment influencing the associated marsh biota. The chemical composition of the water column and the salt marsh varies temporally over intervals ranging from seconds to years, depending on the chemical compounds and the processes under study. Tidal cycles, atmospheric deposition, and riverine inputs of freshwater greatly influence chemical dynamics. Recently the importance of atmospheric exchange with salt-marsh estuaries has been emphasized (10). Previously this type of exchange was poorly understood, and exchanges between the water and the marsh were thought to be the major (or only) exchange pathway for materials.

Oxygen concentrations of the water column associated with salt marshes vary greatly, ranging from anoxia (absence of oxygen) to supersaturated levels. The photosynthetic activity of the vegetation results in the addition of oxygen, and, to a lesser extent, oxygen is added by diffusion from the atmosphere, tidal exchanges, and turbulence of the water. Typically, surface waters are saturated with oxygen throughout the year, while oxygen concentrations at various depths depend on vertical mixing. In many regions, one of the indicators of water quality is oxygen concentration, with low values often associated with waters of poor quality and subject to regulatory remediation. However, even in salt marshes unaffected by human perturbation, dissolved oxygen concentration may drop below presently acceptable regulatory limits, especially at low tide and during periods of low photosynthetic activity. This suggests that low oxygen concentration is a normal phenomenon in certain salt marsh areas and may not always, by itself, be a good indicator of degraded water quality (11).

Important to the functioning of a salt marsh–estuarine ecosystem is the availability and processing of nutrients. The sources of nutrients include runoff from terrestrial systems to river systems and from adjacent highlands, as well as oceanic waters. Nutrients occur both as organic compounds, dissolved organic matter (DOM), dissolved organic carbon (DOC), particulate organic matter (POM), and particulate organic and other substances. These nutrients are utilized and chemically altered by the salt marsh–estuarine biota, resulting in a very dynamic system. The quantity of available nutrients, especially carbon, phosphorous, and nitrogen, influences various biotic components of the ecosystem. In general, nitrogen is important as a potential nutrient which can limit marsh productivity; phosphorous is sometimes limiting, while carbon is almost never limiting. On the other hand, excess nutrient concentrations alter system dynamics by stimulating excess biomass production, especially by primary producers, a process known as eutrophication (see chapter 2).

By virtue of their location where the ocean meets large river systems, salt-marsh estuaries are exposed to other chemical substances, many of which are toxic when concentrations are sufficiently high. Examples include effluents and stormwater inputs carried by upland runoff into rivers or introduced directly into estuaries from human activities associated with estuaries. In a crude sense, estuaries represent the last section of an extensive terrestrial-freshwater–based sewer system which empties into the ocean.

The biology of salt marshes has been extensively studied using various approaches. Taxonomic studies have focused on describing and/or listing the various plant and animal species present. Several of these studies include notation of various species found in southeastern estuaries and their habitat preferences (12–18). Another approach is to study the ecology and life history of a single species. The literature on this subject is too voluminous to be included here. In addition to the references on southeastern estuaries listed above, there are many papers that summarize a representative number of species-oriented studies (19–22). A third approach is to study the physiological adaptations of biota to the environmentally stressful life encountered in the marsh estuarine system (19, 20, 23–27).

Rather than emphasizing the ecology, physiology, or taxonomy of one or a few species, another approach is to view the salt-marsh estuary as an ecosystem and group the various species into broader functional and structural units, such as primary producers, secondary producers, and decomposers.

Obviously there is overlap among these different approaches. For example, a study of primary production of *Spartina alterniflora* could be classified as a study of a single species, but since this species is so dominant in southeastern salt marshes, the results are important to understanding ecosystem dynamics. Because of the interaction and the interconnectedness of the various salt-marsh–estuarine species with the abiotic environment, any attempt to derive sustainable development models for resource utilization management must consider the results of the three general approaches used to study the dynamic coastal environment.

Ecological functions of wetlands

Wetlands are extremely important to the natural functioning of the coastal ecosystem, but they also have numerous uses for humankind. In general the commercial and recreational significance of wetlands can be readily observed and a dollar amount placed on these values. In contrast, wetlands are viewed by many people as miserable, smelly swamps that should be drained and filled so that they can be used for agriculture, housing or some other perceived beneficial societal function. An understanding of the principal ecological functions of this habitat will help provide a more balanced view of its importance.

1) **Habitat and Biodiversity**   The diversity and number of micro-habitats found in wetlands permit many species to live in the wetlands permanently or on a transitory basis. These transitory residents include numerous species of migratory birds and other bird and vertebrate species from adjacent upland habitats that feed on wetland organisms. Other migrants enter the estuarine wetlands from the coastal ocean to spend an essential portion of their life cycle in what can be viewed as a nursery or spawning ground. Although most coastal residents are familiar with some of the more conspicuous permanent residents (such as oysters and clams) and migrants (including ducks, shrimp, and numerous species of fish), hundreds of species of microorganisms and small-sized invertebrates and larval stages of the macro biota live in the wetlands. These "unseen" yet essential species are vital links in maintenance of the complex food webs necessary to sustain the high levels of productivity associated with wetlands.

2) **Productivity**   Wetlands have one of the highest rates of primary production of all major plant habitats (see chapter 2). Much of the biomass produced by wetland plants is not directly utilized by other organisms; instead the wetland plants die and are decomposed by other organisms, producing particles of various size known as detritus. In many wetlands and their associated waters, especially in the southern United States, this detritus and

the microorganisms attached to it are the primary food source for consumers in the ecosystem, and the wetland/water complex is said to be a detrital-based food web. There is a high degree of primary productivity as well as secondary production. This secondary production is reflected in high biomasses and large populations of permanent residents such as oysters and clams, but considerable amounts of this secondary production are exported when transitory species migrate to the open ocean or to upland habitats.

3) **Removal of pollutants**   Numerous types of chemicals, many of which are toxic, are filtered out by wetland organisms, especially plants, as water moves from uplands and rivers toward the sea. Wetlands are essential in improving water quality by removing toxic substances, and they also are important in removing nitrogen and phosphorus which, in excess, result in eutrophication of coastal waters. These two nutrients are utilized by plants to make more plant biomass, which in turn can take up more chemical substances and when eaten provide energy to higher trophic levels. Numerous studies have demonstrated that wetlands are capable of processing organic wastes, thus functioning as a water treatment facility.

4) **Denitrification**   Some scientists who study the nitrogen balance in our environment have emphasized that, as a result of human activities, more nitrogen is being fixed than is being denitrified by the natural world, resulting in an imbalance in the nitrogen cycle.

5) **Coastal Habitat Stability**   Wetland grasses, such as the dominant salt marsh cordgrass (*Spartina*), have extensive root systems that trap sediments and provide physical stability to the wetlands. This stability plays an important role in flood protection by storing water that would overflow stream banks. The wetlands can also absorb the force and energy of storm surges resulting from high wind and tidal action. In this way wetlands provide invaluable protection to adjacent uplands during storm events such as hurricanes.

Human uses of wetlands

Throughout recorded history, wetlands have been used by humans for a number of purposes. Because of the importance of wetlands as a spawning, nursery, or feeding ground for many commercially important species of fish and shellfish, commercial and recreational fishers and hunters utilize the wetland area. Both commercial and recreational activities have had an important economic impact on coastal communities by stimulating local businesses associated with these activities. Wetlands have also been dredged and filled to create new dry lands for various purposes, including new roads, housing, and dredge-spoil disposal sites (see chapter 7). However, in recent years stricter regulatory guidelines have reduced these activities. In some regions of the United States, especially in the Gulf Coast, levees and deep navigation channels have been constructed to assist in flood control. This activity can result in sediment being swept out to sea instead of being deposited on the surface of wetlands as is essential to their preservation. Upstream dams and reservoirs also impede sediment and freshwater downstream transport, negatively impacting on the structural integrity and biology of wetlands. Further, the removal of oil, gas, and minerals from wetlands can directly destroy habitats and can indirectly result in their demise by subsidence. Although the economic and environmental value of wetlands has been demonstrated only recently, since colonial times nearly 116 million acres (about the size of California) of wetlands has been lost and cannot be replaced. The economic value of such a loss is so huge it would be difficult to attach a dollar value.

In recent years, greater attention has been focused on preserving both saline and freshwater wetlands. As a result of governmental regulations and continuing citizen concern, new concepts have emerged concerning the utilization of wetlands based on compatible economic and environmental principles. See chapter 7 for discussion of the application of these concepts.

Outwelling

Although an artificial boundary line can be drawn between an estuary and the near-shore oceanic waters, there is a rhythmic interchange between these two habitats as a result of tidal action. Since seawater flows in and brackish water out with the tide and with riverine input of freshwater, questions have been raised as to whether the estuary removes nutrients and other substances from the seawater and whether the marsh produces and exports material (a process called *outwelling*) vital to the productivity of the near-shore ocean. Various indirect and direct methods have been used in an attempt to answer this question (28). One indirect method is to develop production and consumption budgets for a marsh estuarine system. Any imbalance is due to either export or import of material. A direct method is to establish transect(s) across the interface between the estuary and the inshore waters and to measure the flux of materials during a number of tidal cycles and at different seasons of the year.

One of the most comprehensive studies utilizing this direct method involved the North Inlet estuary located near Georgetown, South Carolina (29). A few results of this study are presented in table 1. Some constituents are exported and others imported seasonally, and both physical and biological processes were identified as potential factors controlling flux magnitude.

TABLE 3.1

Seasonal material exchange between an estuary, North Inlet, South Carolina, and the Atlantic Ocean, based on data of Dame et al. (29).

(+ = export to ocean: – = import to estuary)

	Winter	Spring	Summer	Fall	Year
Carbon (C) Flux					
Total Sediments	–	+	+	–	+
Particulate Organic C	+	+	+	+	+
Dissolved Organic C	+	+	+	+	+
Phytoplankton	+	+	–	–	+
Zooplankton	+	+	–	–	–
Macrodetritus	+	+	+	+	+
Nitrogen (N) Flux					
Total N	+	+	+	+	+
Ammonia	+	+	+	+	+
Phytoplankton	+	+	–	–	+
Zooplankton	+	+	–	–	–
Macrodetritus	+	+	+	+	+
Phosphorus (P) Flux					
Total N	+	+	+	+	+
Phytoplankton	+	+	–	–	+
Zooplankton	+	+	–	–	–
Macrodetritus	+	+	+	+	+

Although most of the living materials represent a relatively small net flux of all the chemical substances, these living entities are important components of the estuarine/wetland ecosystem. Chlorophyll-a is imported in the summer and fall, whereas ammonium is exported during the same periods, suggesting a feedback loop between marsh estuarine systems and the ocean. Waterborne phytoplankton is consumed and mineralized within the marsh estuarine system, with the resulting nitrogen and phosphorus returning to the ocean as ammonium and orthophosphate. Thus nutrients, bacteria, fungi, dissolved inorganic nitrogen, and detritus are continually exported from this estuary, suggesting that large quantities of organic material are being transformed within North Inlet (30). On an annual basis, North Inlet exports most types of material to the coastal ocean and imports only a few constituents. This net export of material contributes to the metabolic activities of near-shore oceanic organisms.

Based on various studies it is apparent that estuaries need to be protected from abuse by human interventions because of their high productivity, which may also have a positive impact on the productivity of coastal waters.

Mangroves

Dominating the intertidal landscape of shores of tropical and subtropical estuaries, embayments, and tidal lagoons is found the characteristic mangrove community (plate 6, following p. 48). On a global scale, mangrove ecosystems cover about 15 million hectares: 6.9 million hectares in the Indo-Pacific region, 4.1 million hectares in South and Central America and the Caribbean, and 3.5 million hectares in Africa (31). Although mangroves are estimated to fringe about 60 to 75 percent of tropical coastlines, they are restricted to warmer waters of the United States, and are especially abundant in Florida.

Mangrove forests represent one of the few regions where various terrestrial plants grow in full-strength seawater. They consist of several different species of shrubs and trees that have a well-developed root system, termed prop roots. Associated with the prop roots are mats of algae and various species of invertebrates and fish. Because of the architecture of the prop roots and associated algae, tidal water movement is impeded, resulting in the trapping of sediments. This in turn can build a new coastline and islands. During times of storms, the mangroves tend to ameliorate the erosional ravages of wave action, thus protecting shorelines. Prop roots are thought to play an important ecological role in that they can penetrate the anaerobic substratum, facilitating the mineral recycling vital to maintaining the high level of primary production occurring in mangroves. This high level of primary productivity influences the energy utilization of other organisms in the food web within the boundaries of the mangrove; energy is also exported to help sustain neighboring ecosystems. Other ecological values of mangroves should also be considered in managing this ecosystem for the benefit of society. Many commercially important species of fish and invertebrates (especially shrimp) utilize mangroves as a nursery ground providing food and protection. Migratory birds also occupy this habitat on a short-term basis for feeding and lodging.

In addition to playing a significant ecological role, the mangrove is valued for commercial reasons. Some principal competing commercial uses of the mangrove ecosystem include mining and mineral extraction, timber and firewood, conversion to agriculture and aquaculture, impacts from diversion of freshwater inputs, discharge of wastes, domestic and industrial development, and salt pond construction. Unfortunately, as coastal exploitation continues at an ever-increasing speed, especially in developing countries, mangrove destruction is reaching alarming proportions (32–34).

Near shore open waters

Extending from the intertidal zone of the open beach and the mouth of estuaries to the open ocean is a large water area. This area can be arbitrarily divided into two major subdivisions: the coastal waters, or neritic zone, and the oceanic zone. The neritic zone extends to the edge of the continental shelf where the slope of the shelf changes, which typically occurs at a depth of about 150 to 200 meters. However, this dividing line is not well defined, and water from one region frequently will invade and mix with the other. For example, the edge of the warm Gulf Stream frequently invades the cooler coastal water in the dynamic region of Cape Hatteras, North Carolina, where it may alter the thermal characteristics of the region, depending on the season of the year. Typically, a greater variation in physical factors occurs in coastal waters than in oceanic waters, but any variation is considerably less than that normally found in estuaries. Variation is particularly pronounced in waters most influenced by runoff from land and freshwater systems. Runoff can result in increased sediments, reduced salinity, and fluctuation in nutrient concentration. The chemical composition of waters near the shore varies more than that of oceanic waters. Another feature differentiating oceanic and coastal waters is the influence of the relatively shallow bottom in coastal areas, which allows greater opportunity for mixing and recycling of nutrients when bottom sediments are disturbed by wave action. Furthermore, the interaction between a rich near-shore benthic fauna and oceanic species occurs more readily because of the bottom's spatial proximity to oceanic waters.

Unlike many estuaries, nearshore and oceanic waters do not usually contain as much suspended matter. This reduction in suspended matter permits light to penetrate to greater depths than it does in estuaries. The photosynthetic activity of phytoplankton is dependent on light and nutrients, while the survival of much of the zooplankton is dependent on phytoplankton as a food source. This increase in light helps significantly in increasing the amount of plankton present. Although planktonic organisms are prominent in coastal waters, they are not uniformly distributed throughout these waters nor do their population sizes remain the same throughout the year but vary with fluctuations in both abiotic and biotic environmental factors.

Coral reefs

Associated with coastal waters are two habitat types that represent an extensive assemblage of prominent plants and animals: coral reefs and kelp beds (plate 7, following p. 48). Coral reefs are widespread throughout the tropics in shallow coastal waters and abound with an extremely diverse fauna and flora (35–38). Approximately eighty million square miles (about twenty-five times the area of the United States) are covered by coral reefs. Unlike trop-ical oceanic waters, which typically have low productivity, coral communities are very productive because of efficient and local recycling of nutrients. Coral reefs, which originate more from biotic rather than from geologic processes, are made up primarily of corals (various species of coelenterates) and calcareous algae. By the very nature of their architecture they provide numerous microhabitats for other reef organisms. A rich resident infauna of motile and sedentary species is present, and many active free-swimming species are periodically attracted to the reef. Environmental conditions nec-essary for reef corals to live and grow are relatively well known. In general, reef-building corals will grow if sufficient light is available and wherever the water temperature remains above 23° to 25°C and rarely falls below 16°C. In addition, a sufficient circulation of water is required to provide nutrients and to remove sediment from the surface of corals. Reef coral does not survive in low-salinity waters; for example, most species exposed to 50 percent seawa-ter will die within twenty-four hours.

In recent times there has been an increased interest in development of tropical regions for commerce, tourism, and urbanization. Unfortunately these developmental activities have not always been planned and pursued with consideration of their potential environmental impacts. As a result, many coral reefs are being exposed to increasing amounts of anthropogenic stresses both on local and regional scales. The main stresses are nutrient enrichment from sewage disposal, sedimentation, oil-related pollution, toxic metals, and thermal pollution. For example, coral reefs are very sensitive to increased sediment in the water because increased sediment load can smother the immovable living coral reefs. Sediments can also restrict light penetration thereby interfering with photosynthetic activity of symbiotic algae living within the coral tissues. Therefore upland activity which results in sediment runoff due to deforestation and topsoil erosion can destroy a reef. Numerous examples can be found in which expanded upland building of resort facilities extensively damaged the neighboring coral reef. Such an effect is ironic, because a healthy coral reef was usually a prime attraction for the establishment of the resort in the first place. Another problem associated with tourism is that damage to a reef results if visitors are not instructed how to protect this fragile resource. Prohibiting such practices as collecting reef organisms or walking on the reef is often necessary.

Kelp beds

Although the marine environment is typically divided into two major divi-sions, benthic (bottom) and pelagic (water), a third unit, *phytal*, has been identified as a large plant association that has its roots in the bottom zone and occupies a distinctive place in the water column. One of the best examples of the phytal zone is the extensive kelp beds found off the California coast (39, 40). In the total energy balance of the sea, kelp beds play a small role, but per unit of space their rate of primary production is comparable to that of salt marshes, coral reefs, and tropical rain forests. For example, the giant kelp (*Macrocystis pyrifera*) is one of fastest growing species of plant (up to 45 centimeters per day) and may reach a length of 200 feet. In addition to having a high rate of primary production, kelp plays a key ecological role by providing a habitat for many species of plants and animals. To cite one example, forty species of fish have been reported from this marine jungle of giant kelp beds.

In one recent case, a controversy arose because sea otters had been practically annihilated by fishermen harvesting fish, resulting in an increase in the sea urchin the sea otters preyed on. In turn the increase in sea urchins resulted in the mass destruction of kelp—because sea urchins graze on newly settled kelp plants and cut adult plants loose. Typically, most of the kelp is broken down and becomes detritus, providing energy for numerous species dependent on the detrital food web. Kelp beds play a key environmental role in the marine environment. In addition, kelp beds and other large seaweeds, which grow in such habitats as shallow, rocky substrata, have been harvested for various commercial uses such as food for humans and other animals, fertilizer, agar, and alginates.

Because of their close proximity to large population centers of Southern California, kelp forests have been subjected to environmental disturbance caused by discharge of sewage into the ocean. For example, the Point Loma kelp forest was decimated during the late 1950s and early 1960s when poorly treated wastes were discharged nearby. One recent study demonstrated the effects on a kelp forest when San Diego's sewage outfall broke (41). Researchers concluded that this spill had an intense, but no lasting, effect.

Upland habitats

The upland habitats of the coastal zone vary greatly in different regions of the United States. In general the principal habitat types are grouped by the type of vegetation which dominates, such as pine forests, hardwood forests, or grasslands. However, some habitats are characterized by other features such as the geographic location, as in the case of coastal dune habitat or upland swamps. As the coastal zone developed over the centuries, many of the earlier "natural" habitats have been replaced by agriculturally related habitats (including cotton fields, corn fields, and truck farms) or structures and facilities related to human habitation.

In summary, coastal zones are indeed a complex system of abiotic and biotic factors interacting over various time scales from seconds to hundreds of years. Numerous studies have been conducted, ranging from how individual organisms respond to external and internal ecological factors to how large ecosystems such as the Chesapeake Bay function over time and in fluctuating environments. However, as is commonly known, coastal environments are subjected to a changing natural environment due to such events as a hurricane or an unusually warm year, and humans also are constantly devising and implementing new ways to alter the coastal environment to suit some perceived need.

The following four chapters deal with significant perturbations to the coastal region resulting from numerous kinds of human activities: development in response to ever-increasing human population; chemical contamination; biological contamination; and associated human interventions. We discuss the environmental impacts of these activities and describe efforts to negate their adverse effects.

4

The Impact of Urbanization

Before World War II large sections of the country were primarily used for agriculture, with a few large urban areas and many small towns serving the surrounding farm population. After the end of World War II, land development patterns changed dramatically to a suburban model. This approach to development emerged as a result of post–WWII urban housing shortages, the availability of capital for home building (through the G.I. loan program), social mobility, inexpensive housing, and certain intellectual and architectural philosophies in vogue between the late 1930s and the mid-1960s (1). Suburbs first emerged as small populated areas on the edges of cities and larger towns, then rapidly expanded farther and farther out. Suburban development utilized a system called zoning, which spatially segregated the elements of culture (such as dwelling, business, manufacturing, recreation) into distinct zones and subzones of consistent or similar use. Thus, unlike traditional European and American communities, which bring people, commerce and services together, zoning separates the places where people live, work, and shop into distinct consistent use units. The growth of the automobile culture reinforced this physical separation, further disrupting the traditional community and resulting in a pattern of development referred to as urban sprawl.

Sprawl development disperses a population over the landscape, in an enlarging, apparently predictable pattern, away from the urban center. As a result, traffic, local budgets, and ecosystems may be thrown into chaos. The need to move infrastructure outward arises quickly and leads to what mathematicians refer to as a positive feedback loop (or in common parlance, a vicious cycle) constantly escalating costs to maintain a continually expanding infrastructure with common impact on natural landscape and resources. However, suburban development patterns continue to occur particularly in the Southeast, and especially in the coastal region. Unfortunately, it seems likely that the escalating pattern of suburban sprawl development will continue. The purpose of this chapter is to explore the changing population structure along the coast and its impact on the environment, the costs of development in response to rapid population growth, and some thoughts on what steps can be taken to minimize the negative effects of population expansion.

Population changes in the coastal zone

When Europeans began arriving on the shores of North America, the landscape had remained relatively unchanged for centuries. Although Native Americans lived in coastal areas long before the first European settlers arrived, only two to three million lived on the entire North American continent, and they had little physical impact on coastal lands (2). The first European settlers were few in number and were primarily farmers, fishers, or tradespeople. As increasing numbers of immigrants arrived, many favored coastal areas for commerce and industry. Water resources were diverted and nearby tidal creeks were used for waste disposal. Marshes were drained and filled, both to gain more land and to control insect pests. An illustration of how southeastern coastal systems have been manipulated can be found by looking at rice production methods during the eighteenth and nineteenth centuries, when many marshes and estuarine systems were used for growing rice. During that time, South Carolina and Georgia were the major rice producing states in the country, and the modifications instituted to grow rice over time destroyed many acres of marshland. It has been estimated that rice production in these coastal waters was at least in part responsible for the fact that less than 50 percent of the marshes remain today than existed when the nation was settled (3).

Over time better opportunities for agriculture and industry developed inland, and the growth of cities and industry slowed along the coast (4). As shown by the 1970 census report, however, the population of the United States began to shift back from the center of the country to the seacoasts during the 1960s (5). This trend continues today, and it is notable that the seven largest metropolitan areas in the country today lie along coastlines (6). In contrast to earlier times, the growth occurring now is due less to agriculture and industry than it is to recreational and residential development, both aspects of urbanization. In industrialized countries worldwide, urbanization is increasing rapidly (7) (figure 1). Projections are that by 2020, approximately 80 percent of the population in industrialized nations will be living in urban areas. Worldwide, the figure is 62 percent. An interesting aspect of this urbanization trend is that globally, population concentrations are increasing most rapidly in coastal areas.

Although growth in coastal areas is and has been increasing in recent years, it has been erratic on the East Coast. A comparison of the 1980 census with the 1990 census shows that the greatest population increases have notably been in Virginia, North Carolina, South Carolina, Georgia, and Florida (8) (figure 2). Where population growth has been significant, development tends to be concentrated along a narrow strip paralleling the ocean.

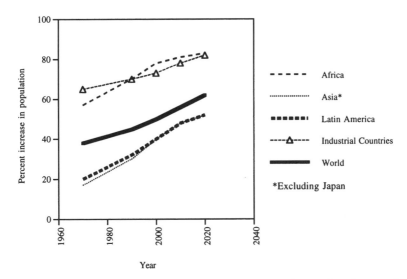

Figure 4.1. **Worldwide view of urbanization in industrialized countries and Latin America** based on data from a United Nations Report (7).

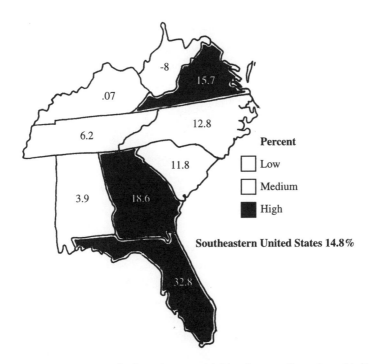

Figure 4.2. **Percent population change within the southeastern United States between 1980 and 1990** (8).

TABLE 4.1

Calculation of population migration and emigration rates of South Carolina (8).

Age Group	Number	Rate	Percent
<5	(8,575)	(3.2)	(7.8)
5–14	(6,367)	(1.2)	(5.8)
15–24	43,733	6.9	37.9
25–34	(39,624)	(6.3)	(35.9)
35–44	17,481	5.0	15.9
45–54	20,738	6.9	18.8
55–64	24,798	8.9	22.5
65–74	26,131	13.8	23.7
75–84	2,084	2.7	1.9
TOTAL	110,219	3.5	100.0

Negative numbers are shown in parentheses.

This pattern of development often presents problems of providing adequate freshwater supply and waste treatment facilities. Furthermore, much coastal development is occurring in relatively pristine and/or rural areas. Those people who are moving toward the coastal areas seek an ability to fish and swim in an environment that is neither degraded or degrading.

Predictions of further growth in the coastal region of the Southeast during the next fifty years vary greatly (6). An increase of as much as 181 percent between 1960 and 2010 has been projected. In the state of South Carolina an increase of approximately 48 percent between 1980 and 2010 has been forecast (9). Recent estimates are that a substantial fraction of the "baby boomer" generation, representing 14 percent of the population, will migrate to the Carolinas and Georgia when they retire. If this estimate proves to be true, then it could signal an even greater population growth than earlier projections. Regardless of which estimate is correct, it is clear that the populations of southeastern coastal states will soar during the next twenty years. The question is not whether growth will occur but whether or not the associated negative impacts to environments and culture can be managed.

Some of the fastest growth in southeastern counties is a result of the increasing number of retirement communities being developed in rural areas. A look at immigration to South Carolina, by age group shows it is highest for those between ages 55 and 74 (8) (table 1). In contrast, the high-

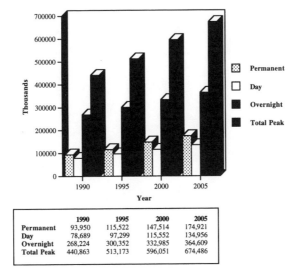

	1990	1995	2000	2005
Permanent	93,950	115,522	147,514	174,921
Day	78,689	97,299	115,552	134,956
Overnight	268,224	300,352	332,985	364,609
Total Peak	440,863	513,173	596,051	674,486

Figure 4.3. **The peak in coastal population of Horry County, South Carolina** (11).

est emigration rates are for those aged 5 to 14 and 25 to 34. Buyers in the coastal retirement communities tend to prefer large lots, as shown in one recent study. As a consequence, 6 percent of the population growth which occurred in Maryland's portion of Chesapeake Bay during a five–year period consumed 65 percent of newly urbanized land. Retirees as well as those seeking a second home are drawn to waterfront lots, and 25 percent of all home purchases are on the waterfront (4). It should be noted that in their desire for privacy and spacious lots, retirees and second-home buyers are no different than 90 percent of individuals in the United States who prefer detached single-family housing (10).

Not only is the number of permanent residents increasing along the southeastern coast, but the number of day and overnight visitors is growing as well. Projections are for total peak population numbers (permanent and day and overnight visitors) to increase significantly in the coming years. For example, in one of the coastal South Carolina counties, Horry County, total peak population numbers are projected to increase from 440,863 in 1990 to 674,486 by 2005 (11) (figure 3). This is one aspect of coastal development that is often ignored.

Impact of rapid population increase

It is human nature to want to change the environment to meet some perceived need. As the population increases along the coast, alterations will

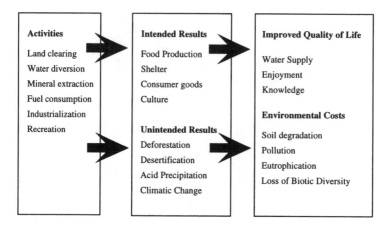

Figure 4.4. **Human activities with unintended ecosystems impacts** (12).

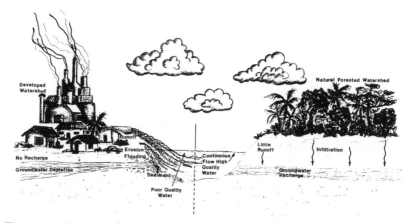

Figure 4.5. **Stream drainage caused by increasing runoff due to residential and industrial development.**

inevitably occur. Often overlooked is the fact that some common human activities can have unintentional negative consequences as shown in figure 4 (12). Forested areas, for example, may be cut down with the objective of industrializing an area and bringing in jobs, improving the quality of life, increasing food production, providing increased recreational opportunities, or constructing houses. But unintended consequences often follow: deforestation, soil degradation, increased runoff, or destruction of natural bodies of water or pollution are but a few examples. The impact on wildlife can be enormous, with the destruction of breeding habitats and the impairment of nursery grounds. Frequently the negative impacts outweigh the positives.

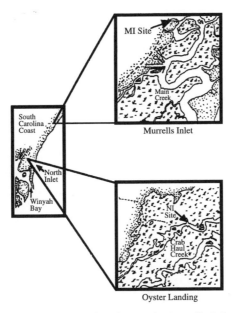

Figure 4.6. **Map showing North Inlet and Murrells Inlet study sites.** Site locations are indicated by arrows (13).

One of the reasons many inland and out-of-state residents find the southeastern coast so attractive is the scenic value of the coastline. With increasing urbanization, however, marked changes can occur in the land-scape. Often forested areas are destroyed and other natural land covers are removed to make way for housing related to urban businesses. Streets and pathways are cleared and/or paved, and the entire watershed is altered. In an undisturbed watershed, runoff is decreased by vegetation that absorbs rain and runoff from the uplands. Rainwater continually recharges the ground-water and flows into rivers or streams, where the water tends to be of high quality. In contrast, if a watershed that has been cleared leaves no naturally covered areas to filter out heavy rainfall, the result is excessive runoff of water (often contaminated with pollutants), erosion, and flooding (figure 5). The excessive runoff of rainwater prevents the recharging of groundwater. Excessive sediment erosion causes river and stream channels to become clogged with sediment and reservoirs to become silted. This process may be accompanied by a decline in water quality.

A comparative study of two estuarine areas along the South Carolina coast, North Inlet and Murrells Inlet (13) (figure 6), demonstrates how destruction of undeveloped land can have an impact on an area. These two regions are about twenty miles apart geographically. Both are bar-built estuaries with similar geological histories. Both are similar in size and are dominated by extensive stands of marsh grass. The primary difference between them is the role of urbanization. The North Inlet area is virtually undeveloped and is surrounded by a heavily forested upland with minor urban development. In contrast, the rapid development along parts of the South Carolina coast is almost shocking, as exemplified by pictures of Murrells Inlet (figures 7, 8 and 9). In 1964 a large part of the area was well forested, with development limited primarily to the coastline. By 1984, housing and roads began to move further into formerly forested areas. By 1997, the Murrells Inlet area had been nearly completely developed, engulfing previously undeveloped areas with fewer trees, more paved roads, and more housing.

The Murrells Inlet area has been subjected to many man-induced stresses resulting from extensive development: restaurants, strip malls, highrises, condominiums and single dwellings, roads, and manipulation of the waterways by dredging, filling, and jetty construction. The impact of development on environmental resources can be illustrated by a comparison of water flow rates. Following a moderate one-hour rainfall, in the North Inlet area the rainwater is absorbed and filtered through the forested uplands. In the comparably-sized urbanized Murrells Inlet area, the initial flow rate following the same one-hour rainfall is twice as high (14). Thus the heavy concentration of human activity in the urbanized watershed has an enormous impact on the coastal aquatic ecosystems and diverts runoff from streets, farms, lawns, residential development, golf courses, and stormwater runoff into the Murrells Inlet estuary.

Another resource that can be negatively influenced by rapid and continuing population growth is the drinking water supply. In coastal plain counties, groundwater provides virtually all the public water supply. Groundwater is found in underground rock formations called aquifers that range in size from thin formations that yield only small quantities of water to large systems that can provide water to millions. In contrast to groundwater, which can be obtained relatively inexpensively from wells, areawide surface water systems require large capital outlays. Many coastal areas rely on groundwater and, as population increases and demand for water grows, water quality problems often develop. If the water level in an aquifer is lowered, salt water intrusion often occurs—as freshwater is removed from the aquifer, salt water moves in. Groundwater quality can also be negatively affected by leaking underground storage tanks, septic tanks, municipal landfills, and agricultural activities. Other factors which have a substantial negative impact on water quality include runoff of fertilizers and pesticides from lawns, dredging activities, and marina development.

Figure 4.7. **Murrells Inlet, 1964.** Note that at this time the area was forested with development limited primarily to the coast line.

Figure 4.8. **Murrells Inlet, 1984.** By this time, housing and roads began to move further into formerly forested areas.

Figure 4.9. **Murrells Inlet, 1997.** By 1997 the Murrells Inlet area had been nearly completely developed, engulfing previously underdeveloped areas with fewer trees, more housing and more paved roads.

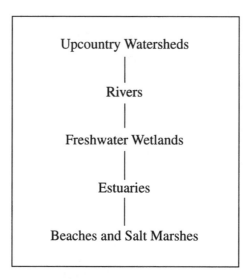

Figure 4.10. **The hierarchical arrangement of landscapes** (16).

As mentioned earlier, one aspect of coastal development that is often not considered is the impact of day and overnight visitors (8). Tourists place significant burdens on community infrastructure, including water, sewer, and roads. Their needs are different than those of the year-round residents, and the growing numbers of summer tourists also have an impact on beaches and other natural resources, including recreational and commercial fisheries. The need to provide accommodations for "summer people" is largely responsible for the development of hotels, condominiums, restaurants, and associated infrastructure. Yet the effects of these transient visitors on the environment are often overlooked until a crisis arises.

Maintenance of the health of a coastal ecosystem depends not only on how the local landscape is utilized but on the management of the whole watershed area. Thus, environmental changes occurring in the upcountry have a profound effect on the watershed of the coastal area, which in turn has an impact on the rivers and estuarine waters. The health of the rivers influences the freshwater wetlands and ultimately estuaries, beaches, and salt marshes (figure 10) (15).

Costs of development

Patterns of development along the southeastern coast are similar to other developing coastal areas where there are widely separated pockets of residential, commercial, and public uses (4). This type of pattern requires per-

Network "law": cost of maintenance (C) increases as a power function, roughly as a square of the number of network services (N). As a city doubles in size, cost of maintenance quadruples.

$$C = N(N-1)/2 \text{ or approximately } N^2$$

A. In early development the major flow of energy must be directed to growth
B. In later development an increasing proportion of available energy must be directed to maintenance and control of disorder

Development

Figure 4.11. **Costs of development** (16).

sonal auto ownership as well as an interlinking road system and puts pressure on local government for increased road construction, often in environmentally sensitive regions. Road construction often poses a question of who bears the cost. Year-round residents have different needs for roads than do summer tourists, and it is not considered reasonable to expect the permanent population to completely fund tourist-generated demand (4).

The costs of development vary over time. A conceptual model of this cost of change is demonstrated in figure 11 (15). This model shows that in early development, most of the energy is directed toward growth. As an area increases in size and complexity, costs escalate to maintain the expanded infrastructure of roads, schools, and water and sewage systems. In turn this means higher taxes for residents. Depletion of natural resources, increased pollution, and degraded water quality can also prove to be costly. Thus unrestrained growth is not an unalloyed economic blessing.

While tourism plays a vital role in the economy of coastal resort areas, it does not ensure prosperity for the total community. For example, in the South Carolina resort areas of Myrtle Beach and the Grand Strand, the number of unemployed and the number of people living in poverty has increased since the 1980 census despite a marked increase in the tourism industry (8). This disparity may result in part because many permanent jobs created by tourism are in the service sector and tend to be low paying. On the other hand, property values do increase, thereby benefitting property owners. The increase in property values usually is accompanied by an increase in taxes that often results in a cultural shift, driving out the poorer residents. However, tourism can provide a long-term solid economic base if the path to *sustainable development* is followed. Sustainable development has been defined by the World Commission on Environment and Development (17) as "development that meets the needs of the present without compromising the ability of future generations to meet their own needs." Achieving sustainable development must include the following initiatives (18):

- In the process of developmental planning, make sure that environmental considerations are not left out.
- Ensure that environmental data are sufficient to implement sound developmental planning.
- Develop public constituencies and make sure the public is well informed.
- Learn what has been done in other areas, both mistakes and successes; to the extent possible use existing environmental and natural resource management capabilities; include everyone who is interested; use a regional planning framework and develop different scenarios for impact assessment.

In summary the Southeast, not formerly exposed to suburbanization to the extent that has occurred elsewhere, has the opportunity to affect development patterns that minimize dispersion of population and infrastructure, conserve open space, and reduce impacts on terrestrial and aquatic resources. Using a slightly higher population density design, landscape architects and planners have begun to relearn how to build "walkable" communities that reduce the need for automobile use and hence the amount of pavement spread on the landscape. This consideration is significant, because creation of impermeable surfaces (paved roads) increases runoff leading to the degradation of receiving water quality (3, 16). One could predict that a reduction in road construction and the development of policies that encourage narrow roadways (as exist throughout the United Kingdom and Europe, and as were common in the United States prior to 1945) could significantly reduce runoff and contamination of waterways. Vegetated buffers are thought to be another part of the solution to the impact of road-induced runoff (19). To date, however, the engineering of buffers—defining buffer dimensions relative to grade, soil type, and acceptable vegetation— is more art than science, and the effectiveness of buffers at ameliorating impacts remains uncertain. Alternatives to road construction (which tends to encourage development and increase traffic congestion) include mass transit, the use of tolls and user fees to pass on the true cost of driving to the consumer, and the building of bicycle and walking paths. All of these measures are linked to development strategies that reduce the need to drive. These strategies include "in-fill" development (developing back toward the cities), reduction in packaging facilities with increased cost for use, mixed-use development (reducing the need to drive and, therefore, the need for roads and road widening), and placing an emphasis on living close to work. Such approaches, when focused on conservation of open space through the use of conservation easements to land trusts, the development of appropriate restrictive codes and covenants and the use of mixed-use zoning and conservation subdivision design (which clusters dwellings to preserve open space), not only preserve natural habitat and reduce the human footprint, but create more liveable communities.

Most conservationists would agree that golf courses are preferable to parking lots. However, typical American golf course designs are environmentally costly (20). Professional and amateur golfing are key pastimes and industries in the Southeast, particularly along the coast. The motivation to redesign golf courses in more environmentally friendly ways (by conserving natural habitats and reducing the impact on resources) is consistent with efforts to manage growth in a manner that leads to sustainable development. The key problems are water, fertilizers, and pesticides. In addition, replacement of native vegetation with sod and exotic species reduces biodiversity and has a negative impact on habitats. New thinking is emerging as far as golf course design is concerned. Substitution of sodded acreage (which requires substantial irrigation, maintenance, and the application of chemical fertilizers and pesticides) with natural and xeriscopic landscaping reduces the human footprint on the environment as well as the cost of course development and upkeep. These so-called "green" golf courses provide aesthetically pleasing, as well as ecologically favorable venues (21). Golf course developers in the Southeast and elsewhere are experimenting with the "green" approach. Ultimately such courses resemble the traditional links of Scotland, rugged and beautiful, though admittedly difficult.

5

Chemical Contamination of the Coastal Zone

Historical overview

Often we tend to think that pollution has been an environmental problem in our coastal areas only in recent times—but history has proved otherwise. Historically, coastal areas have served as magnets for settlement accompanied by changing patterns of use. The world's earliest major cities developed in coastal areas, because of access to trade, transportation, and seafood. At the same time, coastal waters also offered a convenient disposal site for the overflow of unwanted waste. By the fourteenth century, England's Thames River was polluted enough to attract the attention of King Edward III. Riding along the river, he complained that "dung, laystalls, and other filth has accumulated in diverse places in the said city upon the back of the said river." He also "perceived that the fumes and other abominable stenches arising from the corruption would cause, if tolerated, great peril for the persons dwelling within the said city, to the nobles, and to others passing along the river. Fear would arise unless some fitting remedy could be speedily provided" (1). While the polluted Thames not only assaulted the senses, it was also highly unsanitary and unhealthy. Yet only sporadic attempts were made to clean up the river, and then only when it became so foul it was intolerable. Similar conditions were found throughout the world at the time in coastal cities.

With the advent of the industrial revolution in the early 1800s another set of problems arose. Coastal areas, as well as areas located near the riverine systems that emptied into estuarine and coastal regions, became ideal sites for industrial development. A new set of both inorganic and organic chemical pollutants were then introduced over the next 150 years, including metals, polychlorinated biphenyls (PCBs), and polycyclic aromatic hydrocarbons (PAHs) (figure 1).

The period of industrial expansion can be traced over time through the use of cored material from salt marshes (2). Deposits of metals such as lead, cadmium, and mercury began to appear in the early 1830s. Finding their way from various industries into the atmosphere and terrestrial and aquatic systems. Lead contamination is related to smelting activity, and more recently

Historical Trends in Contamination

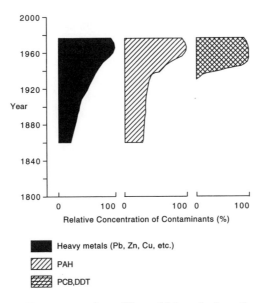

Figure 5.1. **Globally averaged profiles of historical sediment contamination in industrialized countries.** Represents the evolution of contamination in estuarine sediments from industrialized countries over time (years) (2).

to the use of leaded gasoline. Cadmium deposits are found in industrial and municipal effluents. Zinc and copper are found in industrial discharges from mining and smelting, while mercury is given off by coal-burning plants and other industrial facilities. The deposition of metals that began in the early 1800s became more prominent in the 1900s, increasing sharply between 1940 and 1970. This distributional pattern clearly reflects the influence of the industrial revolution. Since the 1970s, levels of many metals have declined as industrial effluents have been regulated and use of lead-free gasoline was mandated. Many metals adhere to suspended particulates in the water column, eventually depositing in the sediment. The observed decrease in metal concentration in upper layers of sediment is largely due to metal-containing sediment being covered with less contaminated sediment.

Organic compounds began appearing in sediment cores around 1930–1940 (2). PAHs are produced by burning of fossil fuels such as petroleum and coal. With increasing automobile use in coastal areas, PAH concentrations rose in estuarine sediment. These compounds found their way into coastal water systems though municipal and industrial discharges, air transport, and runoff from roadways into water systems. They are the only organic contaminants that are now increasing rather than decreasing in coastal sediments (3). Polychlorinated biphenyls first appeared in the early 1930s when these compounds were used in industrial transformers, capacitors, hydraulic fluids, and lubricants. PCBs were banned in 1977, and a slow decrease in the environment has been observed since that time. A number of chlorinated pesticides—DDT, chlordane, and Mirex—reached maximum concentrations between the late 1950s and the mid-1970s. Although these compounds can still be found in U.S. water systems, concentrations have declined over time after they were banned from use.

Contaminants enter the environment either from direct discharge (an identifiable point source) or from contamination from a nonspecific source (non–point-source). Point-source pollution began to receive serious attention in the 1960s. By the early 1970s it became obvious that regulatory action was necessary to control waste disposal into the sea, and in 1972 the Clean Water Act and the Marine Protection Research and Sanctuaries Act were passed. The Clean Water Act regulates the discharge of effluent from industry and municipalities into fresh and marine waters and regulates disposal of dredge material. The Marine Protection Research and Sanctuaries Act regulates the release of sewage sludge and industrial wastes at sea. Since the passage of these federal statutes, levels of point-source pollution from industries has been greatly reduced. On the other hand non–point-source pollution has been increasing (4). Non-point sources of pollution include runoff from farms, construction sites, mining sites, urban cities, and suburban areas as well as excess erosion.

Heavy metals

The sea contains trace amounts of many metals and metal-like compounds (selenium, for example), some of which are essential for normal growth in marine organisms. However, in higher concentrations, certain metals are highly toxic. Discharge of mercury into estuarine and marine waters was recognized as a serious environmental contaminant of marine organisms and humans after serious aliments were attributed to contaminated seafood in the 1940s and 1950s.

The acknowledgment that industrial discharge of mercury was indeed harmful started with the tragedy in Minimata, Japan, when more than a hundred people became permanently disabled and fifty deaths occurred as a result of mercury poisoning (5). This unfortunate event resulted from the bioaccumulation of mercury through the food chain, with the highest levels in top predatory animals (primarily fish), which were then consumed by humans. Subsequently many other studies demonstrated that levels of metals above that found normally in nature could be harmful to marine biota as well as to seafood-eating humans. All metals have certain attributes that can, under certain circumstances, make them particular problems. Many are required in biological processes, and as a result, organisms often have the ability to assimilate them. Further, metals have a number of properties which ensure they will remain within both the aquatic system and the human body. They are immutable and cannot be created or destroyed. Neither can one metal be transformed into another. Metals can be attached to different organic compounds, called *ligands*, altering their various properties. However, the total amount of metal present remains unchanged (5).

Metals and metalloids of particular concern in estuarine and coastal waters include antimony, arsenic, cadmium, chromium, lead, and mercury (4). Sources of pollution come from input of both domestic and industrial wastes. For example, domestic wastewater systems release quantities of copper, lead, and zinc from pipes and tanks. Industrial processes contribute metals from mining and processing of phosphate and other metal ores, finishing and plating of metals, manufacturing of dyes and textiles, and the leaching of metals from antifouling paints used in recreational boating.

As measured by the amount of metals in the sediment, the most highly contaminated areas occur at heavily industrialized sites that have served as chronic disposal sites for sewage sludge and dredged spoil. Examples include Boston Harbor, Long Island Sound, Chesapeake Bay, Baltimore Harbor, San Francisco Bay, Commencement Bay, Elliott Bay (near Tacoma and Seattle), and Santa Monica Bay. The total metal value in these systems is often several times higher than in uncontaminated sediments in nearby areas (6). The contamination of these systems is due primarily to point-source pollution (direct discharge of waste material). As mentioned earlier, since the passage of the Clean Water Act and the Marine Protection Research and Sanctuaries Act, considerable progress has been made in addressing new contamination from point sources.

Non–point-source pollution is much more difficult to control, and only recent updates to the Clean Water Act have began to address this particular issue. One of the increasingly important sources of non–point-source pollution results from urbanization occurring in coastal areas. It has been recognized for some time that urbanization results in significant runoff of chemical contaminants from lawns, road surfaces, parking lots, junkyards and dumps, and that runoff also includes household products and microbial pathogens from septic tanks. Dredging, road construction, and bulkheading lead to sediment input, as do physical modifications of the estuarine habitat.

Few studies have been devoted to estuarine systems where the sole

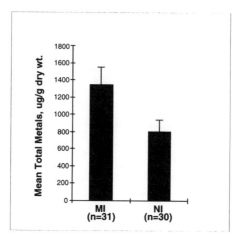

Figure 5.2. **Mean total metals in oysters from Murrells Inlet (MI) and North Inlet (NI).** Murrells Inlet metal levels were significantly higher than those at North Inlet (7).

inputs result from urbanization related activities. Murrells Inlet represents such an estuary. It is a bar-built estuary that has become geared to meet residential and tourist demands. The resident density is 611 /km² during the off-season, but it goes much higher in the summer months due to an influx of tourists. Extensive upland development has occurred, as have dredging, jetty construction, and bulkheading. In contrast, the nearby North Inlet, which is also a bar-built estuary, has extensive marshlands and highlands primarily owned by private foundations and remains largely undeveloped. Using both toxicological and ecological research techniques, scientists are beginning to gain an understanding of the impact of urbanization by comparing these two estuaries. For example, the total level of metals found in oysters from Murrells Inlet was significantly higher than those from North Inlet (figure 2). Since oysters are filter feeders, they accumulate contaminants from the surrounding water. Cadmium, chromium, copper, and nickel were found in oysters at both locations. Lead and tin were found in Murrells Inlet oysters at some locations, but not at all in those from North Inlet. It is likely that higher levels of boat usage and greater runoff in Murrells Inlet contributed to higher total concentrations of metals in oysters. Boat traffic stirs up the sediments, making more metals available in the water column (7). While total metal concentrations were significantly higher in oysters from Murrells Inlet in comparison to those from North Inlet, levels of metal pollutants in both were lower than in oysters found in industrialized estuaries

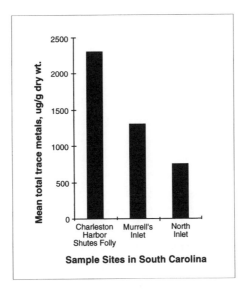

Figure 5.3. **Comparison of total trace metals in oysters from selected South Carolina sites** (7).

(figure 3) . Does this mean that Murrells Inlet oysters are safe for human consumption? Certainly the level of metal contamination is low enough that risks are below levels of concern for individual chemicals. However, cumulative levels of multiple contaminants are relatively high in sites near light commercial activity, roads, marinas, and parking lots. Thus, consumption of Murrells Inlet oysters might not pose acute risks, but it could pose cumulative long-term risks, not only from metals but also from bacterial contaminants.

In marine organisms, heavy metals can interfere with metabolism and health at many levels. Toxicity is often related to a certain stage in the life cycle, with larval stages generally being the most sensitive. Metals occasionally interact with other types of pollutants, such as PCBs and PAHs, compounding toxic effects on populations of marine organisms. However, the level of metals found in small estuarine systems in non-industrialized urbanized areas along the South Carolina coast are much lower than in large urban areas.

Chlorinated hydrocarbons

Chlorinated hydrocarbons or organochlorine compounds are extremely widespread in water, air, and soil as well as in human populations and other animals throughout the world. These compounds include a number of pesticides, herbicides, and fungicides as well as polychlorinated biphenyls used in industrial processes. All have characteristics in common: they are non-biodegradable, they bioaccumulate through the food chain, and they have toxic effects on aquatic and terrestrial organisms.

Organochlorine biocides

Pesticides of one type or another have probably been used since the time of primitive man, but it was only after World War II that organochlorines were used extensively. While many of the pre–WWII compounds used were toxic to humans and animals alike, they were derived from natural substances. With the exception of arsenic, they were not persistent and were generally quickly broken down into harmless compounds once they entered the environment.

The first chlorinated hydrocarbon to be used widely was DDT (dichlorodiphenyl trichloroethane). It was actually synthesized in 1874, but its effectiveness as a pesticide was not recognized until 1939 (6). After 1944, the use of DDT became very widespread; it was assumed to be safe because it killed insect pests, apparently without a negative impact on human populations. Destruction of insect pests increased farm production enormously; thus DDT production and release increased accordingly. In 1947, 124,250,000 pounds of DDT were produced, rising to 637,666,000 pounds in 1960. During the 1950s isolated scientific papers pointed out the potential dangers of DDT, but it was not until Rachael Carson wrote *Silent Spring* in 1962, detailing the harmful effects of DDT to wildlife, that the public became aware of the potential impact of this chemical. During the 1960s, scores of studies demonstrated that populations of some animals, particularly eagles, had been reduced due to DDT and its eggshell-thinning metabolites. Population reductions were reported for crayfish, shrimps, amphipods, annelids, blue crabs, and fish, following single applications of DDT in the range of 0.1 to 0.3 lbs. per acre (8). When DDT was sprayed over large areas of a Canadian forest to kill the spruce budworm *Choristoneura fumiferanca*, the drainage system in the forest lying in the spray zone was contaminated by the insecticide, and many young salmon were killed and stocks reduced to a very low level. When DDT was sprayed, the salmon underyearlings were reduced by 90 to 98 percent compared with unsprayed areas; larger fish were reduced by 70 percent. When the concentration of DDT was cut to half, young fish were reduced by 50 percent and the older ones by about 20 percent. These harmful effects extended thirty or more miles below the spray zone. Four to six months after the spraying, with the onset of cold weather, significant numbers of delayed mortalities occurred among stressed fish that had been in the contaminated water (9).

The effects of direct absorption of pesticides in an ecosystem are fairly

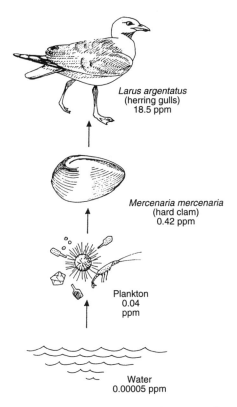

Figure 5.4. **An example of biological magnification of DDT residues** (8).

dramatic; the accumulation of persistent pesticides in food chains is even more so. Analyses of DDT residues in animals in an East Coast estuary revealed not only an increasing concentration of DDT residues as the size of the animal increased, but also greater concentrations of DDT residues in higher carnivores than in those at lower trophic levels (8). Thus the total residues in plankton were 0.04 ppm, while at the highest trophic level, they were 75 ppm in the ring-billed gull. There was twice as much residue in the needlefish as in mudminnows, on which it feeds. It is also of interest that the water contained only 0.00005 ppm of DDT residues, while birds at the top of the food chain had residues about a million times greater (figure 4). Despite the fact that DDT was not applied directly to estuarine and marine waters, this pesticide had found its way into these systems through land runoff, rain, and airborne particulates.

One of the most dramatic effects of DDT was the observed thinning of eggshells in birds, which in turn resulted in dramatic population decreases.

By the early to mid-1960s, the damage DDT had on fish and birds in the environment was widely recognized. Furthermore, many of the target pests had become resistant to DDT. Production of this compound in the United States and in western Europe dropped dramatically and ended altogether in 1972. Countries in the subtropical and tropical regions of the world, however, stepped up production. Although DDT spraying ceased in the United States, its breakdown products, DDE and DDD, continue to be found. Even today, these chemicals often occur in some estuarine sediments at concentrations that exceed sediment guideline values (10). Related DDT-like pesticides include aldrin, dieldrin, chlordane, lindane, endrin, kepone, mirex, and toxaphene. Chlordane was primarily used to control termites; the others were used to control agricultural pests. All are toxic to freshwater and estuarine fish and shellfish. Kepone contaminated the lower James River Estuary in Virginia, in 1975, from manufacturing (6). Mirex was used in the southeastern United States to control fire ants but was banned in 1978 because of its toxic effects on fish, birds, and other wildlife. Toxaphene was widely used by cotton growers in Georgia and Texas in the 1940s.

Residues of chlorinated pesticides can still be found in the environment and in resident marine biota in sites that were at one time heavily contaminated. Often these still-contaminated areas are located near heavily industrialized and urbanized sites, and in estuaries, receiving a large amount of runoff of contaminated water. In Murrells Inlet and North Inlet (a relatively pristine estuary) the only chlorinated pesticide recovered from oysters were breakdown products of chlordane. At both sites levels were very low, and the amount found in oysters from North Inlet was significantly lower than that in Murrells Inlet oysters. The major source, most probably, was chlordane used to treat homes for termites. The fact that oysters still carry residues of this pesticide after use was banned years ago speaks of their persistence in the environment.

PCBs

As with DDT, polychlorinated biphenyls (PCBs) were synthesized in the 1880s, but industrial production did not begin until 1929 (6). PCBs had many commercial applications, particularly in the electric utility industry, where they were used in electrical capacitors and transformers. By the early 1970s, it became obvious that PCBs were being widely spread in the environment, and their toxic effects were having an impact on terrestrial and aquatic animals. By 1977, industrial production was halted. PCBs are still widely distributed throughout the world, although levels continue to decline. Questions also arose about their impact on human populations, especially through consumption of contaminated fish. Even today fish advisories are issued for striped bass taken from New York's Hudson Bay. Relatively small concentrations of PCBs have also been found to interact with other pollutants and increase toxicity to estuarine organisms. Very low levels of PCBs were found in oysters from both Murrells Inlet and North Inlet, and there was no significant difference between oysters taken from the two areas, although one was suburbanized and the other was not.

Endocrine-disrupting chemicals

Residues of chlorinated hydrocarbons and other environmental contaminants have recently been linked to what are termed *endocrine disrupters*. Exposure to small amounts of these chemicals can interact with the endocrine system in humans as well as domestic and wildlife species, resulting in adverse health consequences. As mentioned previously, it was recognized by the 1960s that DDT reduces embryonic survival in birds due to eggshell thinning and cracking (6). More recently, other pesticides have been linked to abnormal sexual development in reptiles and birds, and feminized responses in male fish (11). The pesticides and other chemicals involved are called endocrine disrupting chemicals, or EDCs. Laboratory studies indicate that mixtures of pesticides and certain other environmental chemicals are estrogenic (that is, they mimic the female hormone estrogen) in very small quantities. For example, the pesticides dieldrin, toxaphene, and endosulfan have relatively low potencies alone, but when combined with other environmental contaminants, their potency is greatly enhanced (11). Widely distributed pesticides and other chemicals in the environment reported to have reproductive and endocrine-disrupting effects are shown in table 1 (12).

Oil pollution

When one thinks of oil pollution associated with coastal areas, the first thing that comes to mind is a major oil spill. Yet large oil spills account for only 12 percent of the oil entering the coastal environment each year (13). Another 30 to 50 percent of intact crude and refined oil released into the environment comes from routine operation of coastal oil refineries and installations, including onshore storage and tankers (4). Natural oil seeps on the seafloor, most often in areas where oil production is high, also contribute to oil pollution. Additional oil enters coastal systems from river runoff, urban runoff, transportation activities, and municipal and industrial wastes (6). It has been estimated that approximately one-third of all the oil released into the marine environment ends up in estuarine systems, with seventy-five percent of all accidental spills in the United States occurring principally in estuaries,

TABLE 5.1

Chemicals with widespread distribution in the environment reported to have reproductive and endocrine-disrupting effects (12).

Herbicides	Fungicides	Insecticides	Nematocides	Industrial Chemicals
2,4-D	Benomyl	ßHCH	Aldicarb	Cadmium
2,4,5-T	Hexachlorobenzene	Carbaryl	DBCP	Doxin (2,3,7,8-TCDD)
Alachlor	Mancozeb	Chlordane		Lead
Amitrole	Maneb	Dicofol		Mercury
Atrazine	Metiram-complex	Dieldrin		PBBs
Metribuzin	Tributyl tin	DDT and metabolites		PCBs
Nitrofen	Zineb	Endosulfan		Pentachlorophenol (PCP)
Trifluralin	Ziram	Heptachlor and		Penta-tononylphenols
		H-exposide		Phthalates
		Lindane (Y-HCH)		Styrenes
		Methomyl		
		Methoxychlor		
		Mirex		
		Oxychlordane		
		Parathion		
		Synthetic pyrethroids		
		Toxaphene		
		Transnomachlor		

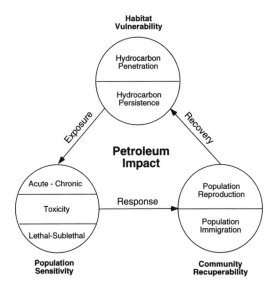

Figure 5.5. **Schematic representation of the impact of petroleum hydro-carbons on marine environments** (15).

enclosed bays, and wetlands (14). A schematic drawing illustrating the impact of petroleum hydrocarbons on marine environments is shown in figure 5. The presence of oil in a fisheries area may make fish and shellfish unpalatable to the consumer and have a negative impact on recreational or commercial fishing (16). The most visible impact of many oil spills is on seabirds, especially those that live on the surface of the water and those that dive to collect food.

How organisms respond to oil pollution depends on a number of factors. The type of oil and the amount available to the organism is important. A number of studies have demonstrated that crude oil and lightly refined products are less detrimental than highly refined products (4). The degree of weathering of the oil before it reaches the estuarine biota is also a factor. That is, oil that has been in the environment for long periods of time tends to evaporate and decompose to a certain extent before reaching estuarine habitats, and is less toxic than near-shore fresh oil contamination. Certain habitats are more vulnerable to oil pollution than others. High-energy environments, like rocky shores and fine-sand beaches exposed to strong wind and tidal action, are less vulnerable than sheltered tidal flats and marshes such as those commonly found in southeastern estuaries. Salt marshes are among the most vulnerable of all, since these environments represent the final depository of oil released into the marine environment. Salt-marsh fauna are also quite sensitive to oil toxicity. Length of exposure is a critical factor, as is the depth of oil penetration into the sediment. If the oil sinks deep enough into the sediment it can accumulate and persist for years (4). The stage of an organism's life cycle when exposure occurs is also important, with developing stages being generally more sensitive that adult ones. Environmental variables (for example, salinity and temperature, which can fluctuate widely in estuarine systems) affect toxicity of oil. Impacts of oil contamination increase especially during periods of low salinity and high temperature regimes. Thus, the season of the year can influence the impact of an oil spill on exposed animals.

Some groups of organisms are more sensitive to oil pollution than others. Among the *meiofauna*, the very small organisms that are near the base of the food web, tiny shrimplike copepods are much more affected than are worms in sediments, such as nematodes (16). Very small, free-swimming organisms called plankton that live in the upper layers of the ocean are also negatively affected, as are bottom-dwelling benthic communities inhabiting subtidal and intertidal habits. In contrast, larger animals are relatively resistant to the effects of crude oil but are much more affected by highly refined oil (4). Fish and motile shellfish, such as crabs and shrimp, can avoid oil spills, but sessile organisms (oyster, clams, and mussels) are more susceptible since they cannot flee the insult or stop taking in water.

Many attempts have been made to lessen the effects of oil spills, including containment of the area with barriers, actually removing the oil, or sometimes increasing dispersement throughout the environment by using chemicals and burning and cutting to remove affected vegetation (4). None of these methods have been very successful, and some have had a more negative impact than the oil itself. When the oil tanker *Torrey Canyon* was wrecked off the coast of Cornwall in 1967, more than 2 million gallons of detergent were dumped into the sea to disperse the oil pollution. The detergents proved to be toxic to shellfish, including the limpets and barnacles on the rocks along the beaches. These animals, which not only act as scavengers but also help to control seaweed growth, were killed in large numbers. As a result of the high mortality rate of these animals, the rocks and the beaches became green with heavy growths of seaweed. In areas where oil, but not detergent, reached them, there was no great mortality among the limpets and barnacles and the area showed a much more rapid recovery (17). Thus a large impact on estuarine and salt marsh biota results from major oil spills. The major oil spills reported to have had the greatest impact on salt-marsh biota are shown in table 2 (4).

TABLE 5.2
Oil spills reported to have impacted salt marsh biota (4).

Tanker/Date	Date	Location	Oil Type	Reported Impacts
Torrey Canyon	March 1967	English Channel	Kuwait Crude Oil	Salt marsh plant species survived all but the heaviest contamination; chlorotic symptoms in some plants. Recovery apparent within 10 months.
Florida	Sept. 1969	Buzzards Bay Massachusetts	#2 Fuel Oil	Plants, crustaceans, fish and birds suffered high mortality immediately after spill. Recovery not complete after 7 years. After 20 years, traces of biodegraded oil were still evident at some sites, resulting in cytochrome P4501A induction in *Fundulus*.
Arrow	Feb. 1970	Chedabucto Bay, Novia Scotia	Bunder C Fuel Oil	Initial effects of oil included smothering of fauna and extensive mortality of *Spartina alterniflora*. After 5 years, intertidal oil not static, but continuously released. After 20 years, a full range of weather oil residues persist.
Brouchard 65	Oct. 1974	Buzzards Bay, Massachusetts	#2 Fuel Oil	After 3 years, marsh grass unable to be reestablished in lower intertidal. Reduced numbers and species of interstitial fauna. Affected areas had higher erosion rates.
Exxon Refinery Spill	Jan. 1990	Arthur Kill, New Jersey	#2 Fuel Oil	Fiddler crabs (*U. pugnax*) from contaminated salt marshes salt marshes demonstrated abnormal behavior.

Polycyclic aromatic hydrocarbons

Among the most ubiquitous contaminants found throughout the world are the polycyclic aromatic hydrocarbons (PAHs). Historically natural background concentrations occurred as early as the 1800s; present-day sediments are now as much as forty times greater than prior to the 1800s (2). Unlike many of the other organic contaminants, PAHs are increasing rather than declining in coastal sediments (3). The PAHs are anthropogenic compounds resulting from combustion-related processes and from petroleum discharges. They enter coastal waters from many sources, including stormwater runoff, industrial discharges, outboard engines, atmospheric deposition, and oil spills. Some or a few of the PAHs are also among the most carcinogenic compounds found in estuarine systems. PAH concentrations reflect urban development more than trace metals (7). Studies have shown that urban runoff is the primary contributor of PAHs to some estuaries (18), and the specific PAHs found in seawater are similar to the PAH compounds found in urban air (19).

A wide range of PAH compounds are found in estuarine waters, some of which are more toxic than others. In sediments from the urbanized area of Murrells Inlet, three of the more toxic compounds—phenanthrene, anthracene, and fluoranthene—were found at all thirty stations sampled. PAH distributions in Murrells Inlet sediments were uneven, however, with the patterns of heaviest concentrations correlated with a number of factors. There was a strong relationship between PAH concentrations found in runoff and the size of nearby parking lots. Statistics were also dependent on how long cars were parked, and the distance to an estuary (20). Traffic intensity and average speed driven were also factors in increased runoff concentrations. At the stations sampled, some of the PAHs found in Murrells Inlet were not found in sediments in the relatively undeveloped North Inlet, and overall total PAH levels in sediments from North Inlet were significantly less than in Murrells Inlet. Still, sediment levels in Murrells Inlet were lower than the more highly urbanized and industrialized Charleston Harbor area (figure 6).

Concentrations of PAHs in oysters were also significantly higher in Murrells Inlet than North Inlet (figure 7). The PAH of highest concentration in Murrells Inlet oysters was the highly toxic fluoranthene. As with sediments, PAH concentrations in oysters varied markedly in Murrells Inlet, and the distance from the nearest paved road was highly correlated with concentrations of PAHs found in Murrells Inlet oysters (21).

While PAHs are generally acutely toxic at very high levels, many sublethal effects on marine organisms have been reported at environmentally realistic concentrations. Skin lesions, cataracts, and liver diseases have been

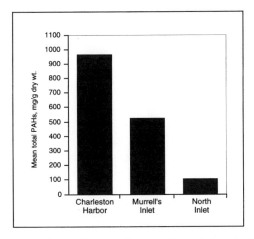

Figure 5.6. **Total PAHs in sediments from selected South Carolina sites** (7).

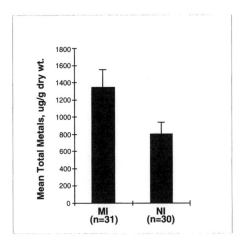

Figure 5.7. **Mean total PAHs in oysters from Murrells Inlet (MI) and North Inlet (NI).** The mean total PAHs at Murrells Inlet was significantly higher than at North Inlet (7).

found in bottom-dwelling fish (4). Normal morphological development of molluscan bivalve embryos was affected when the organisms were exposed to PAH-containing sediment elutriates (22). Some marine organisms can significantly bioaccumulate PAHs. The sediment-dwelling meiobenthic polychaete *Streblospio benedicti* can bioaccumulate sediment-associated fluoranthene at high levels over a twenty-eight–day period. This polychaete serves as a major food source for bottom-feeding fish, and thus can serve to move PAHs through food webs (23).

A number of studies have shown that PAHs interact with trace metals and chlorinated pesticides to produce toxic effects in marine organisms. Cumulative levels of PAHs in Murrells Inlet were not high enough to cause acute toxic effects, but levels were high enough to pose chronic risks to marine biota. Levels of PAHs in Murrells Inlet near sites of light commercial activity, roads, and parking lots can also pose potential human health problems for those who consume clams and oysters from such areas.

Assessment of environmental contamination

How does one determine if chemicals in the sediments and the water are available to organisms and, if so, are they toxic to them? The National Status and Trends Program (NS&T) administered by the National Oceanic and Atmospheric Administration (NOAA) evaluates possible toxic effects of contaminants in the environment. These tests use three different species of organisms with different sensitivities, and measure both chronic and acute lethality (24).

- A sediment-living amphipod is utilized to determine the percent survival when placed in dirty sediment.
- Larvae or gametes of an invertebrate species—often embryos of sea urchins—are exposed to water extracted from dirty sediment and the percent survival and percent normal development are determined.
- Reduction in bioluminescence by bacteria exposed to water component, extracted from dirty sediments, is measured (Microtox TM test).

These three tests not only measure overall degrees of toxicity but permit comparisons of sampling stations and spatial patterns in a given aquatic system. This approach is thus used to determine areas of contamination within an estuary, since toxicity often may be restricted to certain hot spots in highly industrialized areas. For example, in Charleston Harbor, 43 percent of the areas sampled were toxic to the most sensitive test organism; 28 percent were toxic to the next most sensitive (the sea urchin eggs); but none of the areas tested were toxic to the least sensitive test organism (amphipods). The amphipod tests are the least sensitive of these three, but when significant mortality does occur, the structure of the benthic population communities can be greatly affected. The spatial extent of sediment toxicity from nine estuarine systems is shown in figure 8.

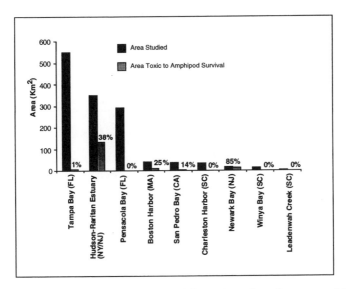

Figure 5.8. **A comparison of the spatial extent of sediment toxicity in nine survey areas** (24).

Toxic effects on marine/estuarine species are often due to the interaction of several contaminants. For example, normal morphological development of bivalve embryos has been shown to be strongly correlated to the presence of a combination of PAHs and tributyl tin (22). Trace metals, PAHs, chlorinated pesticides, and PCBs as well as nontoxic components of the environments such as organic carbon and fine grained sediment were found to be correlated with toxicity in amphipods and Microtox TM tests (24). Thus evaluation of the health of a system must consider the contaminants that test organisms encounter in their environment.

Sediment tests do not address the impacts of contaminants on all stages in the life cycle of an organism. Bioassays have recently been designed to provide information on life history and multiple generation studies of meiobenthic organisms (25). These invertebrates provide a food source for juvenile fish as well as many juvenile and adult benthic-dwelling invertebrates. Since meiobenthic animals complete their life cycles within the sediment, they are exposed to any toxicants associated with sediments. Generally these sediment-associated toxicants are the most persistent and abundant contaminants in estuarine ecosystems. Meiofauna may have ten to fourteen generations per year, and their reproductive output is quantifiable, making them ideal animals to test the impact of contaminants throughout the life

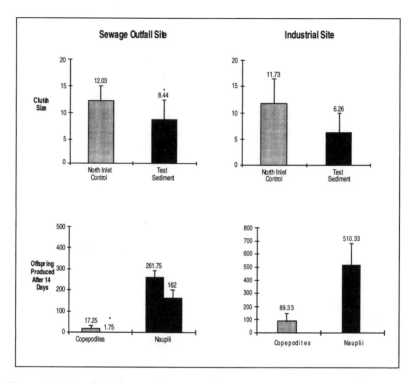

Figure 5.9. **Results of a 14–day reproductive bioassay of the amphipod** *Amphiascus tenuiremis* in contaminated muddy sediments from an organic-contaminated sewage outfall site and a metal-contaminated industrial outfall site (18).

cycle (18). In one set of experiments reproduction was measured for a copepod, *Amphiascus tenuiremis*, in field-contaminated sediments (mercury and cadmium) from an industrialized site, and effluent from a sewage outfall site that contained PAHs, other organic chemicals, and some metals. The sewage outfall site was highly toxic to reproduction (figure 9). Survival of larvae and juveniles reared in the metal-contaminated sediment from the industrialized site was unaffected, but the number of eggs produced was significantly depressed, indicating that over the long run, the population size would decrease. This suggests that both sites could not support copepod populations. Since the meiobenthos are a critical part of the food chain in marine and coastal environments, their survival is a key link in the health of the ecosystem.

As mentioned earlier, the main group of chemical contaminants usually associated with urbanization are the PAHs. To examine the susceptibility of

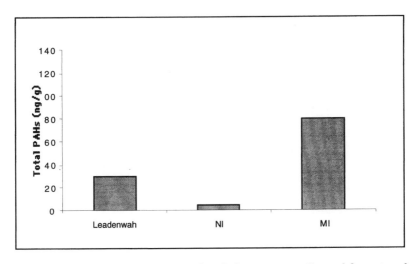

Figure 5.10. **Mean total tissue PH levels in oysters collected from Leadenwah Creek and those deployed at the North Inlet and Murrells Inlet sites.** Clearly, North Inlet is significantly less contaminated than the other two estuarine areas (7).

organisms to urban pollution, oysters were taken from a pristine site, Leadenwah Creek, a small body of water located south of Charleston, South Carolina. These oysters had PAH levels lower than those of Murrells Inlet (a developed estuary) but higher than those from North Inlet. When Leadenwah Creek oysters were translocated to the undeveloped North Inlet estuary, the level of PAHs in oysters decreased to a significantly lower level than they contained initially. In Murrells Inlet, however, levels of PAHs increased significantly (figure 10). In other studies, copepods were exposed throughout their life cycle to contaminated sediments found near a Murrells Inlet trailer park. The molting efficiency of the copepods as they transitioned from larval to adult stage was reduced as much as 900 percent (25). Sediments from this area contained relatively high metal and PAH concentrations. Larger adult estuarine animals (such as grass shrimp and mummichogs) showed no excess mortality when exposed to Murrells Inlet water, although juvenile stages of the sheepshead minnow were negatively affected. Although survival of adult grass shrimp maintained in Murrells Inlet water was unaffected, striking differences were observed between densities of grass shrimp in the two areas (figure 11). Over a sixteen–month period, the average density at Murrells Inlet averaged 2,611 per 16.3 meters of stream. At North Inlet, this figure was 75,157 per 16.3 meters of stream. In other words, there was a 97 percent reduction in grass

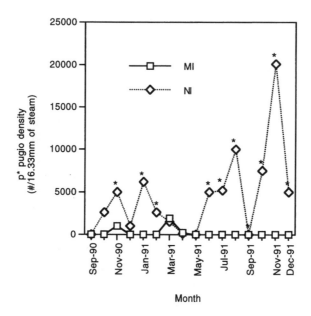

Figure 5.11. **Grass shrimp density measured by block seining at Murrells Inlet and North Inlet sites.** Asterisks indicate significant site differences (25).

shrimp densities at Murrells Inlet in comparison to that in North Inlet. These differences are probably due to a combination of factors. First, the physicochemical environment has changed over the years at Murrells Inlet, primarily due to dredging and filling activities so that it is no longer as dynamic as North Inlet. Salinity fluctuations and dissolved oxygen concentrations are much more pronounced in North Inlet. Since grass shrimp have a greater tolerance to both low salinity and low dissolved oxygen than many of their predators, the wider fluctuations of these factors may be protective of grass shrimp in this type of environment. Another important factor is the reduced stands of marsh grass at Murrells Inlet, for marsh grass is a preferred habitat of grass shrimp. Water quality also decreased more in Murrells Inlet and, while not toxic enough to produce high moralities in the adult grass shrimp, the reduced population of grass shrimp in Murrells Inlet is probably the result of the combined effects of physical habitat modification and chemical contamination (25).

In summary, of the chemical contaminants found in estuarine and coastal area sediments, PAHs are the most closely correlated with urbanization. Other chemical contaminants, although often still considerably higher than background levels, have shown a slow but continuing decline since the late 1970s and early 1980s. In contrast, sediment PAH levels continue to increase. Studies have shown that this increase is related to adjacent land use, including distance to boat ramps, parking lots and light commercial activity, paved roads, and nearby housing density.

Generally levels of chemical contaminants in edible oyster tissue are lower in oysters taken from underdeveloped estuaries than in those harvested in industrialized/urbanized areas. However, oysters from an urbanized estuary along the South Carolina coast were found to have relatively high levels of PAHs, especially those in close proximity to urbanized areas. While levels of chemical contaminants in oyster tissue are lower in urbanized estuaries compared to industrialized ones (and low enough that human health risks are below levels of concern for individual chemicals), cumulatively they do pose chronic risk to seafood consumers. This is particularly true for shellfish harvested near urbanized areas. Thus the harvest of sedentary shellfish for human consumption should be restricted near these sites.

6

Biological Contamination of the Coastal Zone

While chemical contaminants remain a problem in coastal waters, the amount of contamination has, for the most part, been declining since the early 1980s (1). The main exceptions are the polycyclic aromatic hydrocarbons, often referred to as PAHs, and chemotherapeutants, which are on the increase. PAHs, which are produced by combustion-related processes and from petroleum discharges, are discussed in chapter 5.

Generally speaking, levels of chemical contaminants in coastal waters do not pose acute risks, although they can be a cumulative chronic risk to seafood consumers and to marine organisms. But these risks are usually predictable and preventable. In contrast to chemical contaminants, biological contaminants have increased markedly over the past two decades, and they do pose acute risks to human consumers as well as to fish, shellfish, and marine mammals. Not enough is yet known about predicting or preventing the risk of biological contaminants, although progress is being made. These biological contaminants may be toxic and include single-celled algae or phytoplankton, macroalgae, bacteria, and viruses. This chapter addresses the impact of biological contaminants on human and ecosystem health.

Harmful algal blooms (HABs)

Phytoplankton are single-cell plants that form a major part of the base of aquatic food chains. These organisms are the primary producers of aquatic systems, converting carbon dioxide and water into organic matter in the presence of sunlight through a process called photosynthesis (see chapter 2). There are thousands of phytoplankton species, most of which are nontoxic. A few dozen species, however, are highly toxic and the number of occurrences of outbreaks of toxic algal blooms has increased dramatically over the past two decades. Scientists term these toxic species *harmful algal blooms* or HABs.

HABs have been reported for many years, but usually these events were relatively isolated and occurred in restricted areas. Recently, however, almost all coastal areas of the world are affected, and the incidence of HABs is escalating (2). Outbreaks of HABs in the United States before and after 1972 are shown in figure 1. Why have HABs increased so dramatically in recent

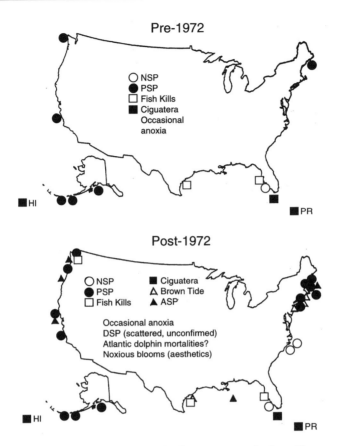

Figure 6.1. **HAB outbreaks known before 1972 and after** (3).

years? There is consensus that the observed increases are real, but there is no agreement as to why. A number of reasons have been put forth, some of which include natural mechanisms such as currents or storms, human activities that enhance nutrient enrichment of coastal waters, expansion of aquaculture activities accompanied by enrichment of coastal waters, unintended transport of toxic species through ballast water in ships, and long-term climatic changes (2).

The impact of the observed increases in HAB outbreaks has been devastating. Mass mortalities of wild and farmed fish and shellfish, mammals, and seabirds as well as increased incidences of human illness and death from eating contaminated fish or shellfish have resulted from HAB outbreaks. Some HABs, but not all, produce toxins that are liberated when the algae is eaten (2). The toxins in these HABs are described as *paralytic, diarrhetic, neu-*

rotoxic seafood or *amnesic shellfish poisoning* (PSP, DSP, NSP and ASP, respectively) (2, 3). Another related problem is *ciguatera fish poisoning*, or CFP, which results when fish consume toxic algae attached to seaweed. Found primarily in tropical and subtropical fish, CFP is widespread and accumulates through the food chain. Since CFP is fat-soluble, it is readily stored in fish tissues, making the largest and oldest fish the most dangerous to eat. It has been estimated that as many as 50,000 people worldwide are affected annually. Other HAB species can kill fish without toxins through the use of spines with serrated edges that cause death to fish by lodging in their gill tissues (2).

Two examples of HABs are red tides and a relatively recently discovered small phytoplankton species, *Pfiesteria piscicida*, which has been named the "phantom killer" because of its many life stages and elusive habits.

Red tide

When toxic phytoplankton undergo rapid multiplication, usually through asexual reproduction, they are said to form a *bloom*. During a bloom microalgae species dominate the phytoplankton. Since the algae are often pigmented, the color of the sea may change from its usual color to red, green, or brown when the algae undergo rapid proliferation (2). The two most common toxic red tides in the United States are NSP in the southeast and PSP in the northeast and northwest. The change in color can be very striking, and early observers termed some of the blooms "red tides." One of the complicating factors in using color as diagnostic tool, however, is that nontoxic phytoplankton also have blooms and, since some are pigmented, they can change the color of the water. Further, some toxic algae are present in dilute quantities and are only noticed when a fish kill occurs.

One of the most intriguing questions concerns the movement of red tides into new areas. Some phytoplankton blooms can be transported over very large distances. For example, a toxic algal species (*Gymnodinium breve* or NSP-causing algae) is usually associated with Florida's west coast, but in the late 1980s it appeared in the water off North Carolina. Based on satellite images of sea-surface temperature, scientists believe that this toxic algal species originated a thousand kilometers away from North Carolina, off the southern coast of Florida (4). It appears that the bloom started from the Gulf of Mexico and was then transported to the southeastern coast of the United States by current systems, ending up in the Gulf Stream. Then water carrying the phytoplankton from the Gulf Stream moved on to the North Carolina coast, where the toxic algal species was discovered (4). Fortunately, the ocean current pattern causing this was unique, and this species did not reoccur in North Carolina.

Other species do reoccur, often on an annual basis. Some are able to form thick-walled, dormant resting cysts and settle on the seafloor, waiting until favorable growth conditions return. A cyst is a small, resting stage of microorganisms. When conditions again become optimal, the organisms excyst and undergo rapid proliferation. The cyst stage offers an excellent strategy for dispersal over wide areas, and when currents carry a bloom of algae capable of forming cysts, the species can readily colonize new areas. One of the most spectacular red tides began appearing in 1972, reaching from Maine to Massachusetts, following a September hurricane. This red tide has recurred annually for two decades (5)

The environmental causes of red tides are undoubtedly many, but there does appear to be a strong correlation between the number of red tides and increasing coastal pollution. For example, the population in the watershed around Tolo Harbor, Hong Kong, grew sixfold between 1976 and 1986 while red tide events increased eightfold (6). Increased nutrient supply from pollution accompanying human population growth would appear to be an underlying mechanism in this case. Since nutrients are essential to phytoplankton growth, nutrient-rich discharges into coastal waters from industrial, agricultural, and domestic waste are suspect. A coastal current of freshwater or low-salinity water has also been identified in some red tide events (7). It is unclear if such waters offer nutrients needed for rapid growth, but it is known that many species of toxic algae require high concentrations of specific nutrients before they can proliferate. Furthermore, the production of toxins by some algae is clearly influenced by specific nutrients (3). At least some of the required nutrients are phosphorous and nitrogen, both of which are often associated with polluted discharges.

The impact of red tides can be measured in terms of human health as well as the health of marine animals, destruction of fish and shellfish fisheries, and economic loss. For example, when toxic phytoplankton are filtered from water by clams, mussels, oysters, or scallops, the shellfish accumulate toxins to a degree that can be lethal to humans or other sea animals consuming them. When outbreaks of toxic poisoning do occur, fish and shellfish beds are closed, resulting in a negative impression of an area and major economic loss to the tourist industry and commercial fisheries in addition to the negative impact on both human and marine animal health.

HAB outbreaks can have a devastating effect on fisheries. Recurrent brown tide outbreaks in New York State beginning in 1985 had an impact on the bay scallop industry for a number of years. Economic losses in this case were estimated at $2 million per year (8). In a single PSP (paralytic seafood poisoning) incident in Maine, cost to the state was estimated at $7 million.

Outbreaks of toxic blooms have become severe enough in some states to require shellfish-monitoring programs to detect algal toxins, at an annual cost of approximately $1 million per year. Clearly the causes of the increase in toxic blooms must be addressed if coastal areas are to continue to be attractive to human populations.

The phantom HAB, Pfiesteria piscicida

In 1991, scientists were culturing the tropical fish *Tilapia* in the laboratory. After several days of exposure to freshly collected water from the Pamlico River in North Carolina, the fish died (9). Puzzled by this unexpected die-off of fish, researchers repeated the procedure and found large numbers of a small phytoplankton species, called dinoflagellates, in the water just before the fish died. Many phytoplankton species which are associated with blooms and fish kills are dinoflagellates. These small organisms are motile and are characterized by two tails called flagellae. After the fish died, a sharp decline was observed in the dinoflagellate numbers unless fresh fish were added to the water. If no fresh fish were added, the dinoflagellates transformed into an amoeboid form or resting cysts and settled to the bottom of the aquarium. Suspecting that the dinoflagellates might be involved in some of the unexplained fish kills in the waters off the North Carolina coast, researchers sampled Pamlico River water during a major fish kill of approximately one million Atlantic menhaden. Upon examining the water in the area of the kill, they found large numbers of a small dinoflagellate, *Pfiesteria piscicida*, during the first day of the kill, but, as noted in the laboratory, very few were observed the day following the die-off (10).

Pfiesteria has a complex life cycle. The majority of the stages are amoeboid in shape or encysted in a resting stage in bottom sediments when fish or shellfish are not present in numbers. The flagellate stage is transitory. When schools of fish or shellfish move into range of *Pfiesteria*, the amoeboid stages and bottom-dwelling cysts transform into the flagellated stage, the most toxic of all the stages. The flagellates release an endotoxin(s) that stuns the fish, and then the flagellates proceed to eat the fish tissues and blood cells. The fish epidermis sloughs off and causes open ulcerative lesions to form. These toxic outbreaks last less than twenty-four hours (10). Since the toxic flagellate stage transforms into nontoxic amoeboid stages or encysts and sinks into the sediments after the death of the fish, *Pfiesteria* is easily overlooked as the causative agent in a fish kill. This species has proven lethal to every fish and or shellfish species tested (12), and, of fifteen major fish kills along coastal North Carolina from May 1991 through November 1993, *Pfiesteria* was identified as the cause of over half (10). In addition to its impact on fish and shellfish, *Pfiesteria* has also been shown to impact human health through skin contact or inhalation of aerosols from contaminated water. Symptoms in humans include eye irritation, respiratory effects, skin lesions, short-term memory loss, and compromise of the immune system (10).

Since the discovery of *P. piscicida* in the early 1990s, scientists have learned a great deal about environmental conditions optimal for toxic activity. The species is found predominately in estuarine waters during the warmer months of the year. Although found in water temperatures ranging from 6°to 33°C, the most favorable temperature is 26°C or higher. The optimum salinity is 15 parts per thousand, or about half the salinity of full-strength seawater (11). Waters that receive nutrients from wastewater treatment plants, fish-processing plants, domestic animal farms, or phosphate mines are the sites of the most frequent toxic outbreaks. Phosphorous enrichment strongly stimulates toxic outbreaks, although nitrate enrichment can also be a factor (9). While *Pfiesteria* has been identified from estuaries along the eastern and Gulf coasts of the United States from Delaware to Florida and Alabama, recent evidence suggests that this organism may often be benign. Using molecular probes, water samples were tested for *Pfiesteria* from 170 sites along the East Coast from New York to northern Florida (figure 2) (13). Only twenty percent of the samples tested positive. In some sites fish kills had occurred due to *Pfiesteria*, but this species was also found in sites where there was no historical evidence of fish kill events. Whether this commonly found species turns into a toxic form or remains benign suggests that environmental conditions are a key factor (13).

Macroalgae

While phytoplankton blooms tend to be short-lived, macroalgae or seaweed blooms can last in the environment for years to decades unless the nutrient supply decreases. Many of the macro algae are nuisance species with extensive filamentous growth patterns that outcompete and overgrow sea grass beds and coral reefs. As is true of phytoplankton, macro algal blooms have been increasing in recent years, especially along coastlines with increasing development. Almost always the increased growth is in response to nutrient enrichment from the addition of phosphorous and nitrogen from agricultural activities as well as increased amounts of domestic wastewater and sewage. These nuisance macroalgal blooms have contributed to the decline of sea grass beds that are used as nurseries for fish and shellfish. In Florida, macroalgal blooms have been implicated in the drastic declines observed in commercial fisheries in recent years with resultant economic loss. Many coral reefs off the Florida coast are being overgrown by harmful macroalgal

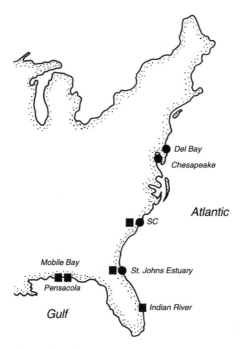

Figure 6.2. **Detection of *Pfiesteria piscicida*** in estuarine waters along the eastern coast (13).

species, which will in time destroy the reefs along with tourist-related industries (3).

Bacteria and viruses

It has long been recognized that disease-causing bacteria and viruses can contaminate fresh and estuarine waters and, subsequently, fish, shellfish, and humans. Today this problem has become a major concern in many coastal areas. Pathogens are associated with agricultural and urban runoff, malfunctioning septic tanks or sewage treatment plants, and storm and sewer overflows as well as overboard toilet discharges from small or recreational boats. The greatest concentrations of bacterial contamination lie at the interface of the land and sea (14, 15) and can be linked directly to upland population as well as to the number and concentration of marinas (16). In coastal regions, human exposure to pathogenic organisms can occur from eating contaminated shellfish, swimming and playing water sports, or drinking freshwater from contaminated wells. When estuarine pollution occurs, shellfish beds are closed or restricted and the number of beach closures increases. Tourism-based communities particularly can economically suffer heavily when swimming beaches are closed during the prime tourist season. In the following section let us consider further the causes and implications of disease-causing bacteria and viruses.

Causes of contamination

One of the major problems facing those living in the burgeoning coastal areas is that of human waste disposal. In many areas where rapid development is occurring, especially in formerly rural areas, the necessary infrastructure is not in place to handle increased amounts of both discharge water and sewage. Consequently, in areas lacking sewage treatment plants, septic tanks become the method of choice for waste disposal in newly urbanized areas.

While septic tanks can be a safe and effective method for waste disposal in some areas, much of the southeastern coastal plain is flat and the water table high. Indeed, the water table can rise to just inches below the soil surface in some places, especially during heavy seasonal rainfall. During these periods effluent from septic tanks can rise to intersect septic tank disposal trenches, and, as a result, effluents from septic tanks are discharged directly into the groundwater. Other studies also indicate that failing septic tanks are the primary source of surface water bacterial contamination in some areas (15–19).

Even in areas with sewage treatment plants, population growth and urbanization can have an impact on water quality downstream if used water and treated sewage exceed the capacity of treatment plants. For example, as much as 1335 million gallons of treated sewage is discharged into the Chesapeake Bay watershed per day, resulting in lower estuarine water quality. This, in turn, leads to closure of shellfish harvesting areas and illness to swimmers. In fact, the number of shellfish harvesting areas in the southeast placed off limits due to bacterial pollution increases almost annually (20).

In coastal areas there is a clear correlation between upland agricultural practices, increased population density, bacterial contamination and closure of shellfish harvesting areas. Filter-feeding shellfish, such as oysters and clams, which readily grow in contaminated estuarine water, can take up significant numbers of bacteria. As a result, over 30 percent of the shellfish harvesting areas in the United States have been ruled as unsafe for harvest (16). The link between population growth and increases in bacterial levels can be well illustrated by comparing the amount of fecal coliform densities in estuarine surface water in a developed area (Murrells Inlet) versus an undeveloped one (North Inlet). Measurements were made in these two areas during

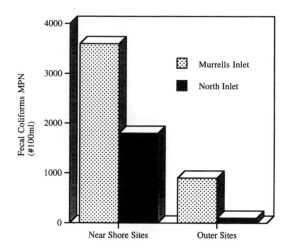

Figure 6.3. **Spatial differences in fecal coliform densities in Murrells Inlet and North Inlet.** Significant differences were noted at the near shore land-estuarine interface and the outer beach-sea interface. At Murrells Inlet both the land-estuarine and the outer beach site (Garden City Beach) receive run-off from populated areas (15).

the summer and fall, the peak period for large numbers of tourists. The greatest densities of high fecal coliform densities were found near the highly urbanized areas of Murrells Inlet and Garden City beaches (figure 3). Coastal marinas have also been shown to cause significantly higher coliform densities in surface water. The combination of septic tank runoff and marinas are largely responsible for the fact that two-thirds of Murrells Inlet sampling stations exceeded the Water Quality Criteria Standard for fecal coliform bacteria compared to one-third of North Inlet stations (15).

Impact on human health and shellfish

In a study of pathogenic bacteria along the South Carolina coast, increased bacterial infections were found among residents who consumed ground and/or surface water located near a septic tank field (14). Increased infections were also noted in people who swam in contaminated estuaries or who ate shellfish taken from such areas. The most commonly found pathogenic bacteria along the South Carolina coast were *Campylobacter, Salmonella, Shigella,* and *Vibrio. Campylobacter* infections cause diarrhea, fever, nausea, and sometimes vomiting. *Salmonella* species can cause infections that range from simple gastroenteritis to severe enteric illnesses. Species of *Shigella* cause bacillary dysentery but rarely cause other types of infection. The *Vibrio* species cause a range of diseases, the most serious being *V. vulnificus*, which can cause fulminating septicemia and the group of cholera strains that cause pandemic cholera. Greater than ninety percent of seafood-borne *V. vulnificus* infections result from the consumption of raw or undercooked oysters. Other species of *Vibrio* are associated with human wound infections. All of these species are principally sewage-associated pathogens affecting inshore and coastal fish and shellfish (14).

Outbreaks of gastroenteritis are associated with two viruses, Norwalk and Hepatitis A. Both involve sewage-contaminated oysters and clams. Since these animals are filter-feeders, they can become contaminated if they pump virus-contaminated water through their gills and deposit the virus within their gut and edible tissues. Seafood may also become contaminated from poor hygienic practices or from use of nonpotable water during processing and/or handling. Viruses are difficult to eliminate. They enter coastal waters via sewage effluents, and current practices for chlorination are generally ineffective. Furthermore, viruses may persist in water, seawater, and sediments for over a month (21).

The evidence of the negative impact of cholera on human health has been well documented for many years. Now there is evidence that oysters growing in polluted waters also can be negatively affected by the cholera-causing *Vibrio*, especially in oysters also infected by the oyster parasite *Perkinsus marinus*. This parasite, which tends to infect oysters from polluted, high-salinity estuaries, is suspected of affecting the immune system of the oysters, making them more susceptible to cholera-causing *Vibrio* infections. The increases in both the oyster parasite and *Vibrio* can be linked to the impact of urbanization. Urbanization tends to reduce the inflow of freshwater into estuarine areas, resulting in high-salinity waters, which predispose the oyster to parasitic infections. Thus, if the waters are also polluted, the result is a greater accumulation of *Vibrio* in oyster tissues. Human consumption of oysters grown under such environmental conditions increases the risk of *Vibrio* infections (15).

Measurement of bacterial contamination

For many years indicators and standards of sanitation in the United States have been measured by sampling densities of total and fecal coliform bacteria present in 100 milliliters of water, using techniques adopted by the American Public Health Association (24). Using these methods the most probable number (MPN) of total and fecal coliform bacteria can be obtained. On the whole these techniques have been quite successful, and the United States has the safest drinking water and food supply in the world (22).

However, the traditional total and fecal coliform indicators of contamination are not always perfect. Some viruses can be present in water samples when there is no evidence of bacterial indicators. For example, chlorination of waste water kills most if not all fecal coliforms, while untreated and undisinfected waste water shows high fecal coliform levels (23). Yet, the level of viral infections is essentially the same in untreated as in treated waste water (24). Since a number of outbreaks of shellfish-borne illnesses among human shellfish consumers have been linked to viruses, it would be of great value to have a standard that indicates when viruses pose a threat to human health. Another area where fecal coliforms do not serve as a good indicator of infectious agents can be found around offshore wastewater discharges, since fecal coliforms generally fail to survive in offshore waters (25).

In recent years, questions have been raised about the relative significance of non–point-source pollution from animals versus humans. As humans have taken over more and more coastal lands, birds and other wild animals have been forced into others areas where there is less competition between humans and wildlife. As a result, in some coastal areas where human population is limited, total fecal coliforms from birds and other animals can be high enough to make harvesting of shellfish unacceptable. Yet, fecal coliforms from nonhuman sources generally do not pose a threat to human health. These observations raise questions about the relative significance of non–point-source pollution from wild animals versus that from humans. In an extensive study of Cow Trap Lakes, Texas, it was found that the sources of contamination in that body of water primarily came from a large migratory bird population in the watershed. An analysis of both oyster and water coliforms determined that the bacteria present came from nonhuman sources and presented no health threats to humans (23).

It is possible to distinguish between human and animal pollution. Using a technique called pulsed field gel electrophoresis, the DNA of *E. coli* from animals and humans can be determined. Using this technique on samples of water from a watershed in Virginia, it was found that raccoons were the major cause for shellfish harvesting closures in these waters (25). It is estimated that 60 percent of the time the pulsed field gel electrophoresis techniques enabled scientists to identify whether or not pollution sources were from animals or humans. Another technique uses profiling the fatty acids of *E. coli* from animals and humans. This technique allows discrimination between human and animal fecal contamination 90 percent of the time (26). It is also of interest to note that *E. coli* from humans has a greater resistance to antibiotics than that of wildlife. Measuring antibiotic resistance gives an 80 percent probability of discriminating animal versus human sources (27).

These methods, however, are still in the experimental stage and have not been developed yet to the point where they can be performed on a routine basis as easily as the now standard MPN technique. With the increasing pressure on our coastal water resources, efforts must be intensified to ensure that shellfish beds are closed due to human fecal coliforms, but not to non-human ones (23). Furthermore, methods must be developed for measurement of viral contamination in both water supply and in shellfish to protect human health.

In summary, biological contaminants have increased markedly over the past decade. Harmful algal blooms (HABs) now affect almost all coastal areas of the world. While the exact causes of the dramatic increases of HABs are not fully known, some of the more likely causes include nutrient enrichment of coastal water, runoff from land use (agricultural activities, phosphate mining expansion of aquaculture and disposal of wastewater). Outbreaks of HABs have had an impact on wild and farmed fish, and shellfish, mammals, and seabirds as well as increased incidences of human illness and death from eating contaminated fish or shellfish.

Urbanization and population growth in coastal areas have been linked to increased outbreaks of pathogenic bacteria and viruses. As a result of these outbreaks, more and more shellfish-harvesting areas have been closed and an increase in outbreaks of human illnesses have been reported. Both land use and wastewater disposal have a significant impact on bacterial and viral contamination.

7

The Impacts of Other Human Interventions

As seen in chapters 4, 5, and 6, human society has altered coastal environments through developmental and industrial activities and by the introduction of chemical and biological contaminants. However, numerous other activities also can influence the ecological dynamics of the coastal ecosystem. This chapter will examine the environmental consequences of some of these other human interventions, including physical changes to the coastal environment (such as dredging, channelization, bulkheading, and beach nourishment) and the introduction of exotic and/or non-native species. Although hurricanes and other storms are not the result of human interventions, the impact of these natural disasters on coastal regions can be enormous, especially in highly developed areas. Hence their effect on coastal areas is also considered in this chapter.

Impact of physical alterations
Dredging

As the population has grown in the coastal zone, there has been an ever-increasing trend to modify the area by some segment of society for perceived specialized needs. Often various proposed projects to alter the coastal landscape lead to conflict, especially as competition increases for use of the finite and limited resources of the region. As a result of public concern about these conflicts, new laws and regulations have been promulgated over the years to regulate development, but in many cases, new legislation governing land use policies is indicated.

One area of concern is the potential impacts of dredging (table 1). The need to dredge for recreational and commercial utilization of the various waterways in the coastal zone is well recognized. Riverine input of silt, originating from regions located far from the coast and from runoff from uplands adjacent to coastal rivers and estuaries, can make harbors and other waterway systems non-navigable. As a port develops, pressure mounts to expand its capabilities to handle more and larger ships. The scenario that follows then is to deepen the harbor to greater depths and/or increase the size of the waterways by dredging the adjacent highlands. Both harbors and coastal waterways, including inland waterway systems, such as those located along

TABLE 7.1

Potential impacts of dredging (various sources)

Dredging Activity	Likely Result	Possible Outcome
Dredging coastal waterway for recreational and commercial uses	Riverine input of silt from uplands adjacent to coastal rivers and estuaries	Increased input of silt can make harbors and other waterways non-navigable
Movement of dredge spoil to another site, following dredging	Shortage of land for disposal; if spoil is contaminated environment rendered unfit for living organisms	Need for seeking disposal sites at sea or remote upland sites
Dredging in establishing benthic community	Slow rate of recovery (3–18 months)	Failure of dominant vegetation to recolonize
In any site, changes in water circulation patters and sediment transport	Changes in penetration of salt water up river systems	Limitation of distribution of wetland organisms, from resultant salinity
Offsite effects of dredging (i.e., dredge disposal)	Release of pollutants into water; increased turbidity	Change of non-distributed site into less productive site

the East Coast of the United States, need to be dredged to maintain sufficient depths for navigation. Until the practice was regulated more stringently and greatly reduced in recent years, extensive dredging of channels and filling projects occurred in freshwater and saltwater marshes and mudflats to permit property owners to have a boat at their front door. When the walls of these channels eroded and fell into the channel, thereby causing a decrease in the depth of the channel, one solution was to construct a bulkhead. Large and extensive bulkheads were also constructed on the edge of estuaries where towns and cities were located, especially in those areas with a highly developed shipping industry and associated maritime businesses.

No matter what purpose it serves, one outcome of dredging is that a quantity of earth (commonly called dredge spoil) must be moved from one site to another. Dredging and the construction of channels often have been done for non-navigational purposes. Dredging has been utilized to drain wetlands to control insects, especially mosquitoes, which were either potential carriers of human disease or a nuisance. Also, drainage channels have been utilized to carry effluent discharge resulting from industrial processes to receiving coastal rivers and estuaries. Various laws and regulations exist governing dredging operations (see chapter 8). Applications for a permit to dredge should address the question of environmental impacts and indicate how to avoid or ameliorate the negative impacts.

An obvious environmental impact is that dredging destroys existing benthic communities of plants and animals by removal of bottom sediments. In addition, the removal of these sediments alters the water quality adjacent to the dredge site by releasing sediments into the water. The result is cloudy water which in turn inhibits light penetration needed for photosynthesis by phytoplankton and submerged vegetation. Also, the resuspension of bottom material in the water column may reintroduce chemical contaminants that can alter water chemistry and harm species located both in adjacent regions and distant sites. Depending on the extent of dredging being proposed, deepening a channel can significantly alter the circulation pattern of an estuary, a change that may then negatively influence neighboring habitats. The altered circulation pattern can accelerate sediment deposition in other parts of the shipping channel, resulting in the need for further dredging.

The actual process of dredging and filling has impacts on wetland habitats. Although wetland habitat is destroyed, a new upland habitat is created. Frequently the sediments deposited in the wetlands find their way back into the channels, thereby resulting in the need for additional dredging. Today wetland disposal of dredge spoil is allowed only under extreme circumstances.

If the sediments contain contaminants, the newly created spoil site will not be a desirable site for habitation by humans or other species. Where a specific estuary had been a site of dredging for many years or where the surrounding uplands had been utilized for commercial and residential development, there may be a scarcity of nearby upland disposal sites, since landowners often object to having a spoil site near their property. For example, in a recent case, landowners initially were willing to have dredge spoil disposal on their low-lying land to create more high land. But when an analysis of the sediment of the proposed dredged material revealed the presence of toxic levels of PCBs, opposition quickly arose. To overcome the obstacle created when dredge spoil is not welcome, distant upland disposal sites are sometimes utilized. However, dredging costs increase markedly when sites are located further inland. Another method of disposal is to take the dredge spoil in a ship and dispose of it at sea on the continental shelf. Both distant upland and ocean sites have negative environmental impacts associated with them—destruction of an existing habitat and the potential for the introduction of contaminants. A variation on this theme is to pump dredged material to renourish nearby eroding open beaches. Again, this practice is contraindicated if the material is contaminated or if the sediment is unsuitable for beach enrichment. Since the open beach is subjected to constant wave action, any sediment containing mud or fine-grain material will quickly be washed away.

Although there are numerous problems in locating disposal sites, one of the key questions concerning dredging which is considered in the permitting process is the potential environmental impact of dredging. Obviously dredging has a direct impact on the region and must be considered. In general, recolonization and eventual recovery of the benthic community occurs, but the rate of recovery varies from three months (1) to twelve to eighteen months (2). Disposal of dredge spoil in coastal wetlands can drastically alter the environment to the extent that reestablishment of the dominant vegetation is precluded (3). If the dredge spoil is deposited as a thin layer on the marsh, the effect is less devastating to plant communities. However, the thickness of the deposited layer is difficult to maintain. One study demonstrated that if the layer is thicker than 5 centimeters, only 5 percent of the plant cover existed after one year and species composition of the marsh was altered for many years (4).

Dredging also may have effects on offsite habitats. If the dredging causes extensive changes in circulation patterns, sediment transport processes are altered and the ecosystem dynamics of wetland regions are influenced. Changes in penetration of salt water up river systems can occur restricting the exposure of the biota to salinity. As discussed earlier, salinity is extremely important to wetland and aquatic organisms and low salinity can limit their distribution. For example, thriving oyster, clam, and shrimp populations may decrease in upstream sites that become dominated by freshwater to the point where it is no longer possible to harvest them. Further, changes in waterways can influence the pattern of boat usage. In post-dredging periods, increased boat traffic can cause erosion of the creek or river banks due to wave action, resulting in the need for more dredging.

An obvious offsite effect of dredging is alteration in water quality. Associated with sediments may be pollutants such as pesticides, hydrocarbons, and metals that can be released to the water column or deposited on land along with other dredged material. In addition, ammonia may be released from sediments stimulating phytoplankton production. In turn, phytoplankton biomass may alter such important water qualities as pH, dissolved oxygen level, and biological oxygen demand. Another effect is an increase in turbidity due to added suspended sediments in the water column. Enhanced turbidity can decrease primary production of the phytoplankton and influence the visual acuity of animals, resulting in a decrease in the ability of an organism to avoid predators or to find food.

Marinas and piers

As a consequence of the great influx of people to the coastal region, a growing number of recreational boaters utilize coastal waters. To accommodate the demand for facilities to meet the needs of these boaters, marinas, individual docks, and boat launching sites have been constructed. Because these facilities are best located along shorelines protected from excessive wave action, estuaries and waterways surrounded by salt marshes are sites of choice. However, construction and operational activities associated with these facilities can have adverse environmental effects (5) (figure 1). The principal areas of concern are habitat destruction and the introduction of pollutants. Further, the construction of an individual dock or marina and the resultant increase in boating activities may have relatively small influence on the ecology of the immediate adjacent region. However, the cumulative effects of many marinas, docks, and boats have been shown to result in environmental catastrophes.

Habitat alteration due to dredging associated with construction of a marina and/or docks and creation of an adjacent boat basin can alter both the benthic community and oyster beds (6). Alteration of the highlands by construction of buildings, roads, and motor vehicle parking can influence runoff of surface water and substances into the system. Pollutants may be introduced into the environment by increased boat usage, especially in areas of restricted water circulation, and by activities associated with marinas and piers, including use of anti-fouling paints and disposal of human waste from sanitary systems on boats. The impacts of chemical pollutants are discussed in detail in chapter 5. In addition to these potential negative chemical environmental impacts, boat operation can create turbulence and wakes in the waterways, resulting in shoreline erosion.

Recognizing both societal pressure to construct marinas and docks and the need to protect the environment, the United States Environmental Protection Agency suggested measures to minimize negative impacts (8). These include designing and locating marinas and docks to facilitate flushing and exchange of water; using effective techniques to control runoff, erosion, and spills; making pumpout facilities and slipside wastewater collection manda-

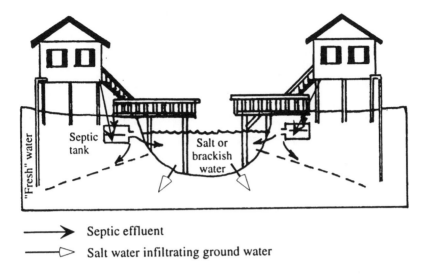

──────▷   Septic effluent
──────▷   Salt water infiltrating ground water

Figure 7.1. **To provide more waterfront lots, finger canals are dredged from existing waterways through marshlands and other habitats to the highlands.** Along the finger canals, boat docks for individual owners or for marinas are built. The principal problems associated with these canals are the lowering of groundwater table; seepage of salt or brackish water into the water table; septic tank seepage into canal and ground water; poor canal water circulation, resulting in stagnation and decreased water quality; nutrient overloading and oxygen depletion; and elevated water temperatures, which may kill organisms (7).

tory; using dredge spoils to create new salt-marsh habitat; and scheduling dredging and construction activities to avoid peak seasons of biological activities (spawning and migration of organisms). Federal and numerous state agencies have set guidelines for obtaining permits to construct and operate marinas (see chapter 8).

Mining and energy development

In some estuarine and benthic coastal areas, extensive mineral deposits and vast oil and gas reserves abound. Mining of such commercially important substances as copper, zinc, tin, titanium, phosphate, gold, and sand and gravel have environmental consequences that must be taken into account when considering permit applications (9). Many of the mining operations involve extraction of the desired substance from open mines. In one type of operation, a small inlet is dammed, the water is pumped out, and then the extraction occurs from an open mine. Obviously the existing habitat is greatly disturbed, but the area may be recolonized by local biota when the mining stops and the dam is removed. Another extractive technique is to use dredges to collect subtidal benthic deposits. This process creates deep holes in the waterways, accompanied by habitat disruption and potential changes in water flow. In some cases, the resultant deep holes filled in with sediment and the benthic region was recolonized by fauna and flora.

The most important nonrenewable resources found in the coastal region are gas and oil. Although many of the oil and gas fields are located offshore, some extraction activities occur in coastal wetlands. A detailed discussion of the environmental impacts of oil pollution is to be found in chapter 5. Extraction of oil and gas from marshlands has resulted in land sinking (subsidence), causing a rise in sea level. Habitat loss due to subsidence is an important concern, as has been observed in Louisiana.

Although not normally thought of as a mining operation, the removal of water from coastal underground aquifers to satisfy the demands of the ever-increasing numbers of coastal inhabitants is creating a crisis in many regions. When freshwater is removed, seawater has been found to invade and occupy the resulting vacated space, a process called *saltwater intrusion*. The decreased water supply then necessitates obtaining freshwater from inland sources or by desalination processes.

Multiple alterations

To judge the environmental impact of a proposed new development or to attempt to implement a plan to correct an existing environmental problem, an assessment take into consideration the potential impacts on other segments of an ecosystem or on other ecosystems that may be distantly located. A decision then can be made on a plan of action based on this assessment. Two problems readily surface and must be addressed: cumulative effects within an ecosystem, and resource utilization conflicts within and between ecosystems. Obviously this assessment is not a simple process and frequently leads to lawsuits.

For various reasons, a request to alter the environment may deal with a relatively small segment of an ecosystem, such as dredging a short boat channel in a marsh. In itself this proposed activity may seem a minor alteration, but when many such requests are considered in a limited area, the potential cumulative effect on the entire marsh ecosystem may result in a major disturbance. Decisions based on one isolated environmental alteration may have dramatic long-term consequences.

Human interventions in the environment of one part of the planet to enhance use of a resource may have deleterious impacts on resource utilization in another distant geographic region. These interventions may represent long-term existing activities that over time have a cumulative negative effect on downstream environments. How such disputes are to be settled (or avoided) is a major problem society must continue to actively work to solve. One example will illustrate this problem.

In the region of the Gulf of Mexico where the waters of the mighty Mississippi River enter the coastal marine waters, an extensive area exists where commercially important marine organisms (as well as other living components of this ecosystem) are almost nonexistent. This area, commonly called the "dead zone," has been growing in size over many years and now covers about 7,700 square miles. Obviously this environmental disturbance has had a disastrous impact on the associated commercial and recreational fisheries based in adjacent coastal states.

Various studies to determine the causes of this disturbance have suggested several potential culprits. An important cause seems to be related to past and present farming practices in the Midwest. Runoff from farmlands, containing nitrogen fertilizers and nitrogen-rich products produced from the manure from hog farms, is carried via waterways to the Mississippi River. The Mississippi system, the largest river system in the United States, drains thirty-one states. The nitrogen-rich input it carries to Gulf waters results in increased growth of algae that are in turn eaten by zooplankton. Bacteria, feeding on feces of zooplankton and on the dead remains of both zooplankton and algae, utilize dissolved oxygen and may reduce the oxygen concentration below levels required for most marine life to survive (see chapter 2 for discussion of hypoxia and anoxia). Coupled with the increased runoff of nitrogen over the years, there has been a steady disappearance of wetlands bordering the streams and rivers—wetlands that function to filter chemical materials from surface runoff. Proposals to restrict the use of fertilizers and/or reduce the amount of farmland would have a serious impact on food production and the survival of many farms.

Another human intervention activity has also been blamed for the creation of the "dead zone." Over the years a tremendous amount of dredging, diking, and channeling has occurred in Louisiana. The cumulative effects of these activities, coupled with the loss of thousands of acres of wetlands, probably have played as important a role in the deterioration of the Gulf as Midwestern agricultural practices.

The controversy surrounding this divisive societal dilemma points out the enormity and complexity of dealing with local and national/international environmental/socioeconomic problems. Local, national, and international agencies must work together to prevent and better deal with complex environmental issues.

Open beaches and beach nourishment

The never-ending struggle between nature and human society over the use of natural resources is graphically represented in the fight to occupy open beaches. The unrelenting forces of the environment to alter a dynamic, unstable region of the earth is pitted against human will to control its environment. This struggle has been going on for centuries but has been more acute in recent times because of the developmental pressure to build structures at the edge of the sea. As documented in chapter 4, intense pressure to accommodate the insatiable desire to be at the beach has stimulated development activities by the private sector and caused state and local governments to adopt measures allowing more and more invasion of the coastal region.

The impact of various environmental forces on this narrow stretch of the earth's landscape between the land and sea, known as the open beach, has been extensively documented over the years (10, 11). This habitat acts as the first line of defense against the excessive wind force of hurricanes and storms. During these natural events, the heightened wave action is absorbed by the beach and dunes, providing a protective barrier to lands further removed from the open waters. However, when dunes and dune vegetation are altered by human construction activities, much of the stabilizing capacity of dunes and their vegetation is lost leaving the open beach vulnerable to damage from unrelenting wave action. Although hurricanes are very dramatic and get a great deal of attention, winter storms, especially northeasters, which occur more frequently than hurricanes, cause extensive beach erosion over time.

Beaches are a very dynamic habitat where opposing forces act either to build up beaches by deposition of sand suspended in ocean water or to erode them when waves remove sand. In general, behavior of sand in near-shore ecosystems must be considered in terms of a sand transport system. This system encompasses a considerable area on both a north-south axis and extending from the beach to several kilometers offshore to depths of thirty to forty feet. Within this broad area, the sand transport system is influenced by various oceanographic factors, including weather conditions ranging from calm to big storms, water currents, and wind direction. To better understand the processes associated with sand beaches, coastal scientists have devised sand budgets that measure inputs and outputs of sand to a beach. The principal input comes from transport of sand unto the beach from coastal long shore

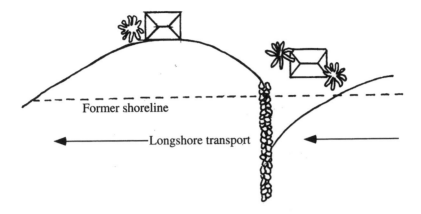

Figure 7.2. **On the updrift side (right) of a groin constructed perpendicular to the shore, sand is trapped, resulting in beach development.** However, the beach on the groin's downdrift side may be deprived of sand, causing the beach to recede and endangering coastal structures near the high-tide mark (7).

currents, from neighboring river systems, and cliff erosion. The outputs of sand from beaches are the result of offshore transport, long shore transport, and wind transportation into dunes (figure 2).

Numerous efforts have been made to stabilize and/or replenish eroding beaches. Landward of open beaches, sand dunes are typically found wherever there is a sand supply and wind to transport it. When windborne sand hits an obstacle, such as vegetation or sand fences, the wind velocity is reduced and sand is deposited and begins to accumulate, eventually building dunes. Although protective during storm tides of short duration, dunes do not prevent persistent day-to-day recession during periods of changes in the shoreline. In regions where the dunes are not protected by law, they frequently are leveled to accommodate the construction of residences and commercial structures such as hotels. Also, visitors to the coast destroy the vegetation by walking on the dunes or by collecting the very distinctive and attractive sea oats, a dominant species that is drought- and heat-tolerant. Those dunes not covered by vegetation migrate depending on winds.

In order to help stabilize dunes, various measures have been used with variable success. Sand fences or other structures that hinder wind velocity help accumulate sand. However, storm surges and intense wind will subvert the beneficial effect of these structures and the dunes will be damaged. People eager to get to the beach frequently establish paths through the dunes, resulting in blowout areas. In many places, beach access is restricted to crosswalks elevated above the dunes. Another remedial technique is to plant vegetation on the dunes. Along the Atlantic and Gulf coasts the principal species used to stabilize dunes are American beach grass, sea oats, bitter panicum, and saltmeadow cordgrass. These species are adapted to the harsh environment of drought, heat, low nutrient supply, wind action, and burial by sand (12). All of these techniques result in temporary improvement and demand constant long-term attention.

Although it is banned or sharply regulated in many parts of the United States, another approach to beach stabilization is to place hard structures such as groins, jetties, and seawalls on the beach and in the adjacent ocean (figure 2). Groins and jetties, typically extending perpendicular to the shore, inhibit the flow of seawater, resulting in sand being deposited on the beach. This procedure may temporarily build up a section of eroding beach, but it deprives beaches downstream from the groin or jetty from being renourished by sand-laden seawater. Hence, one segment of a beach might be stabilized and increased in size at the expense of adjacent areas that are not being renourished by normal processes. In this case, serious objections are raised by landowners who see their beach decreasing in size. By contrast, seawalls are constructed parallel to the beach in the belief that they will function to protect property landward of an existing beach from erosion by wave action. However, a seawall hastens the retreat of the beach in front of it as waves dash against it and erode the sand away. One published study compared the dry beach width on stabilized and unstablized beaches and found it consistently and significantly narrower in front of the seawall (13).

In recent times, beaches have commonly been maintained by replenishment of sand. Sand is replaced on an eroding beach either by pumping or trucking in new sand from another site. Although beach replenishment is a costly and temporary solution to an age-old problem, it continues to be employed, especially in areas where economic investment in adjacent upland is extensive (on beaches immediately adjacent to hotels, for instance). Research is on-going to determine the most cost-effective and environmentally beneficial techniques to continue beach replenishment.

In summary, four general approaches to the problem of eroding beaches are available:

- attempt to stabilize the shoreline with hard structures (i.e., seawalls, groins and jetties);
- attempt to stabilize by use of replenishment of sand;
- plant vegetation or construct sand fences to capture and hold the sand; and

- allow the shoreline to erode or accrete and locate buildings farther away from the water's edge.

The first approach is temporary and may hasten erosion, the second and third are temporary, while the fourth option has been used sparingly.

Introduction of new species

Much attention has been paid to the introduction of foreign and exotic species into a new marine environment. Some species are introduced deliberately, while others are accidental introductions, perhaps carried by ocean currents or associated with shipping activities. For example, ballast water and sediments released from ships while in a foreign harbor are capable of transporting many species from widely different areas. Two studies, one from Hong Kong and the other in the state of Washington, illustrate this point well.

In analyses of ballast water from five container ships that originated from both sides of the Pacific ocean, scientists reported finding eighty-one species from most of the major marine taxonomic groups. Many of these species were found to settle in laboratory culture tanks (14). Not all of the species, of course, could become established, since the viability of some species was affected by the length of time spent in the ballast water. Even those that did survive over relatively long time periods would not necessarily find environmental conditions favorable for growth in a new setting. Still, the potential for the introduction of a new species carried in ballast water does exist.

An examination of ballast water and sediments from bulk cargo carriers of wood chips sailing between Washington State and Japan revealed that many species survived the eleven- to fifteen- day transoceanic voyage. When the sediment was incubated in seawater, a large number of micro algae were found, including diatoms, dinoflagellates and phytoflagellates (see chapter 6). Some of the species in the ballast water were able to form spores or cysts and lived for extended periods of time under unfavorable conditions. Since up to 20,000 metric tons of water and several cubic yards of sediment are discharged each voyage, the possibility of introducing harmful algal blooms would seem to be a genuine threat (15).

Accidentally introduced species are often able to outcompete native organisms, often for reasons not always easily understood. A Mediterranean species of the mussel *Mytilus* was introduced on the South African west coast in the late 1970s and quickly invaded the south and east coasts, where it began to successfully compete with an indigenous species, the mussel *Perna*. In seeking to determine how the introduced species could outcompete the native one, scientists found a high degree of parasitism in *Perna*, but none in the invading *Mytilus*. Many *Perna* were infected with larval stages of a parasitic worm that markedly reduced the hardiness of the mussel both physiologically as well as reproductively. Thus it seems likely that parasitic infections in *Perna* allowed the introduced mussel a competitive advantage over the native species (16).

For an introduced species to successfully compete in a new environment, it is obvious that the invading species must be able to adapt physiologically to habitat conditions, salinity regime, and temperature range. Two species of hydrozoans (jellyfish-like organisms) have successfully established themselves in a number of different regions of the world. These animals are thought to be native to the Black Sea, where low-salinity or brackish water is the norm. In 1992 and 1993 the two hydrozoan species were introduced in the Petaluma River and the Napa River, California. Both of these rivers provided an environment where the salinity varies over the year from 0 to 20 0/00. These two species have become successfully established in the two rivers, and it is believed that these hydrozoans will be able to invade other low-salinity bodies of water in the area over time (17).

Some introduced species have become established and play an important and positive role in the marine ecosystem. In Hawaii a species of seaweed was planted in Kaneohe Bay, Oahu, in 1974. Since that time, the seaweed has spread not only to other places on Oahu but also to the other Hawaiian islands, where it now serves as the main food source of an endangered species, the green turtle, *Chelonia mydas* (18). Other introduced species have been shown to have a profound effect on the environment. In the Chesapeake Bay an introduced submersed aquatic plant, a species of *Hydrilla*, affected community structure, growth of bivalve species, and sediment sand grain size and deposition (19). *Hydrilla* was first reported in Chesapeake Bay in 1984 and was probably introduced via boat traffic. It spread rapidly from the original site, displacing native plants in its path. The number of small fish and some invertebrates increased substantially. The observed increased densities of marine animals were probably due to a number of factors, including an increase in the number of habitats that offered refuge for small organisms, thereby reducing predation pressures. The widespread area covered with *Hydrilla* also can trap particulate matter and provide a food source for animals inhabiting the region. Thus the introduction of *Hydrilla* has had a significant impact on the community structure of the areas where it has become established, by altering species composition. Such major changes further point out the impact of introducing a non-indigenous species into a new habitat (19).

Fisheries and aquaculture

Estuaries have played an important role as a source of food for humans as well as for other species of plants and animals. Commercial and recreational fishers alike have extensively harvested food from coastal waters. At times conflict between these two groups has resulted in verbal and physical expressions of the significant differences in their respective goals. Such topics as the size of nets and the amount of catch per day are examples of their conflicts. One study on the influence of fishing disturbance by towed gears indicated that the magnitude of response varied between species and between habitat types (20). This finding indicates that, when developing local fishery management plans, fluctuations in population size of a food species cannot categorically be attributed to commercial fishing. In some regions unregulated fishing by both groups of fishers has resulted in depletion or diminution of an important food species, a result that should be unacceptable to both.

Estuaries are important not only as a food source of permanent resident species (such as oysters and clams) but as a temporary, but vital, home for about 90 percent of the commercial catch of oceanic finfish (21). Habitat loss or decreasing water quality within estuaries would have dire effects on the productivity of offshore fishing.

Numerous historic examples demonstrate that human intervention (such as overfishing or pollution effects) has depleted or destroyed a fishery. However, fishing remains an important part of our economy, especially at the local level. Attempts have been made at the international, national, state, and local levels of government to manage fisheries activities. For example, the Law of the Sea, an international agreement ratified by many countries, permits a country to regulate its fisheries within two hundred miles of its shoreline. In 1976, the Fisheries Conservation and Management Act was passed into law. It provides for the conservation and management of all fishery resources within two hundred miles of the coast of the United States. Eight Regional Fishery Management Councils were established to develop fishery management plans for their regions. These plans seek to prevent overfishing, while allowing for maximum harvesting of fish based upon scientific information (see chapter 8).

Even with effective management practices it is uncertain whether the sea can provide sufficient fish catch to satisfy world demand. To augment the supply of fisheries products from the sea, mariculture programs have been developed throughout much of the world.

Some forms of mariculture require that large areas of productive waters be fenced off or otherwise restricted for the raising of specific commercial species, such as fish, shrimp, oysters, or clams. This practice has certain legal and environmental complications. Leases must be obtained from governing agencies and security programs devised to protect the areas from poachers. In some temperate zone regions, early life history stages of tropical species of shrimp are introduced into coastal ponds during warmer months of the year to be grown to harvestable size before decreasing fall temperatures kill them. Extreme care must be taken to prevent the accidental release of these introduced species and any accompanying pathogenic species into a new geographical region (see above, in this chapter, on introduction of new species). Other commercial species, including various fish, bivalves, and algae, are grown successfully in ponds.

Various models have been developed to more efficiently increase the yield from these enclosures while protecting the environment. For example, the MOM system (Modeling Ongrowing fish farms Monitoring) has been proposed for intensive marine fish farming. This model should help to maintain healthy environmental conditions in and around fish farms and to provide a useful tool in site selection and coastal zone management (48).

Habitat restoration

The heightened awareness of ecological problems, stemming in large part from awakening environmental activism of the 1960s and 1970s and continuing to the present, has highlighted the massive destruction of coastal habitats over the centuries. This excessive environmental manipulation can best be seen in the case of coastal wetlands. Estimates suggest that approximately 50 percent of these wetlands have been destroyed since the 1700s, with the highest rate of loss occurring in the last thirty years (22). Before being aware of their large-scale environmental and economic benefits, society in general thought that wetlands were at best useless and at worst a dreadful, smelly, disease-ridden hellhole that should be "reclaimed" by draining and filling to create usable uplands. In recent years, laws have greatly curtailed this destructive practice and interest in restoration and creation of wetlands has grown.

Since the smooth cordgrass *Spartina alterniflora* is the dominant species in southeastern salt marshes of the Atlantic and Gulf coasts of the United States, techniques for salt marsh restoration have wisely focused on understanding how this species grows and its role in the ecology of salt marshes. Various methods of introducing *Spartina* into a new area have been employed (5). Direct seeding is the most economical method, but it is only successful in the upper half of the intertidal zone. Transplanting individual plants or plugs, containing roots, substratum and stems has been more successful in lower intertidal areas, but is more labor-intensive. Typically, successful seeding will result in vegetative cover by the end of the first growing season, while transplanting takes at least two growing seasons. Although techniques have been developed for seeding, planting, and transplanting *S. alterniflora*, establishment of this species in new areas will probably be successful only when prescribed environmental conditions are met (23–26).

Once this species is successfully introduced, the question can be raised as to whether or not this new assemblage of *Spartina* is typical of a "normal" functioning salt marsh ecosystem. Although few studies have been undertaken to answer this question (25, 27, 28), some generalizations can be made. Salt marsh habitat creation may eventually result in sites functionally equivalent to natural areas, but it will probably take several decades. The aboveground and the below-ground biomass of cordgrass duplicates that of natural marshes within five years, and the colonization of marsh sediments by macroinvertebrates is relatively rapid. By contrast, the total biomass, abundance, and species composition generally differs from that of natural marshes, probably the result of lower organic matter and changed physical and chemical properties of sediments (29, 30). It is estimated that it would take fifteen to thirty years for levels of organic matter in created marshes to equal that of natural systems; duplication of nutrient pools would take considerably longer (31, 32). Much of the detailed scientific information on salt marsh restoration has been summarized, and the reader is encouraged to consult these papers for a more in-depth discussion (5, 24, 25).

Hurricanes and storms

The coastal zone is also influenced by events originating great distances away and are not due to human interventions, hurricanes, and storms, especially those occurring during the winter that are commonly referred to as "nor'easters." Estimates indicate that the greatest volume of sand located on the tide-influenced coastal ocean shelf is transported during storms. These storms, which occur more often than hurricanes, are notorious for causing extensive open-beach erosion.

The impact of hurricanes along the southeastern and Gulf coasts ranges from destruction of ships caught in the path of a hurricane to destruction of buildings on land, to the death of residents. Hurricanes also frequently create new channels. Hurricanes that affect the eastern coastal regions of the United States and the Gulf Coast, originate off the west coast of Africa and move through the Caribbean. They are ranked according to wind velocity from one to five. A category one hurricane has wind velocities of 74 to 95 miles per hour, a rare category five hurricane has a wind velocity of at least 155 miles per hour.

One of the most detailed studies on the impact of a hurricane on the coastal environment was conducted by scientists at the Belle W. Baruch Laboratory at the University of South Carolina following Hurricane Hugo, an extremely powerful category four hurricane. This hurricane made landfall at Charleston, South Carolina, on September 22, 1989 and moved on up the coast of South Carolina and then inland into North Carolina. The storm struck at high tide and caused massive property damage between Charleston and Myrtle Beach. In the Carolinas fifty hurricane-related deaths were reported, and 129,687 homes were destroyed or damaged (33). Damage estimates included $500 million due to flooding, $1.2 billion in damage to agriculture and forestry, and $4 billion in property damage (34). It is very clear that damage to developed areas was enormous. Immediately following the hurricane, scientists undertook a series of studies to determine the extent of damage to undeveloped barrier islands, marshes, and a low-lying coastal forest. The area selected had, at that time, a twenty-year database of biological, chemical, and physical parameters with which to compare possible changes brought about by Hurricane Hugo. Results showed the impact on undeveloped barrier islands and back barrier marshes to be minimal (35–37). No change could be seen in the salt marsh network of creeks, sandbars, and marsh distribution. However, the coastal forested lands did show damage due to salt stress as a result of a strong storm surge of 15 to 24 feet (37). Two days after the storm the forest floor was covered with water with a salinity of 21 o/oo (38). By January 1990, the foliage of the trees in the area covered by the storm surge was discolored and extensive defoliation had occurred. By June 1991 new growth was flourishing in the salt affected area. Other organisms were also negatively affected, although recovery was relatively fast. Flying insects and birds virtually disappeared immediately following Hugo, but they returned quickly. The numbers of reptiles and amphibians were still significantly lower six months after the storm than populations observed before the storm. Bottom-dwelling invertebrates in the blackwater stream of the forest were also negatively affected, and population density dropped by 97 percent immediately following the storm. However, the animals returned to pre-storm levels in three to six months (36). The pelagic saltwater fauna showed an immediate decrease in population size, but this response was short-lived and the species associated with the shallow waters of the marsh reappeared in numbers within two weeks. The accumulated organic material normally found at the surface of the marsh was washing into and resided in creek bottoms for about two weeks before it was re-deposited on the marsh surface. Intertidal oyster reefs were relatively unscathed by Hurricane Hugo, in sharp contrast to the effects of the more recent Hurricane Bonnie. When this hurricane passed by North Inlet, extreme winds from the west blew most of the water out of North Inlet and the adjacent Winyah Bay, resulting in unusually low seawater levels at high tide. Consequently, for about four tidal cycles, oysters living in the upper reaches of the intertidal zone were not covered by water. This resulted in mass mortality of these animals and an increase in the aroma associated with the region. However, these undeveloped areas, although showing some effects of the storm, were left relatively undisturbed, and those that were affected recovered quickly.

One of the probable reasons Hurricane Hugo had very little ecological or geomorphic impact on marsh invertebrates was that the marsh was protected from the storm surge by three to four meters of water resulting from wind and wave action (39). Another factor was the paucity of rainfall, thereby allowing the salinity to remain relatively stable (37). Hurricanes that are accompanied by excessive amounts of precipitation can result in excessive damage to the sessile organisms in marsh and estuarine environments. For example, in 1955 three storms hit the North Carolina coast, on August 12 and 17 and again on September 17. Following 22 inches of rain in August and 21 inches in September, salinity in the Newport River near Beaufort, North Carolina, dropped dramatically (figure 3). As a result approximately 36 percent of the oysters and 87 percent of the clams were dead within six days after the second hurricane (40). Further mortalities occurred in September up to six weeks after the last hurricane in September. That year damage to the oyster industry in the middle Atlantic states was estimated at $10 million (41).

Unlike the relatively rain-free 1989 Hurricane Hugo, Hurricanes Dennis and Floyd caused extensive flooding to the East Coast of the United States in 1999. The state of North Carolina was especially devastated by the flooding. Approximately 36 inches of rain dropped on the coastal area, and, while this was not as much rain as the 43 inches produced by the three 1955 hurricanes described above, the flooding during these most recent storms was devastating. It was the worst flooding the state had ever experienced, and it was probably exacerbated by past human environmental interventions (49). The release of toxic material, fertilizer, and wastes from the existing hog industry caused extensive human suffering, great economic loss, and environmental damage. Erosion had increased as a result of clearing the land, by cutting trees, plowing the fields, and by the draining and destruction of wetlands that inhibit flooding and filter runoff water. Deposition of pesticides, herbicides, and nutrients (especially nitrogen) arising from fertilizers used on farms and lawns, and livestock wastes and overflowing sewage systems, increased. In a media release, Joe Rudek, a scientist with the Environmental

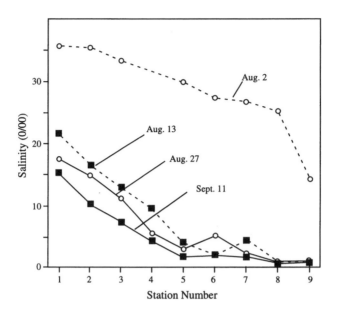

Figure 7.3. **Salinity profiles of the Newport River, North Carolina. August 2, 13, and 27 and September 11, 1955.** Hurricane Connie passed on August 12 and Diane on August 17, 1955 (40).

Defense Fund, estimated that over 100 million gallons of lagoon wastes were spilled, an amount about ten times greater than waste associated with the Exxon Valdez oil spill (50).

Not only was the flood harmful to the inland regions of the North Carolinian coastal flood plain, but much concern has been expressed about the immediate and long-term environmental impacts on the coastal marine waters of Pamlico, Albemarle, and Currituck sounds. The floodwaters emptying into these sounds carried pollutants, organic chemicals, and other substances that run off from uplands, along with the overflow from lagoons and sewage systems. The impact of the chemical pollutants is yet to be determined. Initially shellfishing was prohibited, but it is now permitted. The concentration of some toxic substances in water entering the sounds was elevated, but not to a level of concern. According to Dr. Patricia Tester (51), a scientist at the NOAA laboratory in Beaufort, North Carolina, the marked elevated concentrations of nutrients from fertilizer in floodwaters resulted because water remained standing over farmlands for a long period of time, soaking up fertilizer before exiting to the sounds. This increase in nutrients has raised phytoplankton levels, providing more food for zooplankton. In turn, increased zooplankton populations provide more food to animals higher on the food chain. Long-term effects on the productivity of the sounds will need to be determined by future studies. The great surge of freshwater into the sounds did decrease the salinity to freshwater levels in Pamlico sound. The impact of this "salinity shock" on the survival of organisms, especially the sessile species, is yet to be assessed. The increase in organic material and nutrient content and the increased primary production of phytoplankton will create a favorable environment for increased microbial activity. The net effect could result in a "dead zone" as described earlier in this chapter. Provided adequate research funds are made available, scientists and economists will be able to study and better understand how ecosystems and human society withstand and respond to major ecological insults.

Hurricanes, of course, are not new to the East and Gulf coasts of the United States. They tend to occur in clusters and are known to have occurred since the time of Columbus. Indeed, hurricanes played a major role in the historical development of our country. As early as 1528, Spain tried to establish the first European settlement on the continent near what is now Pensacola. The fleet was wrecked by a hurricane, killing all but ten of the four hundred men sent to settle the area (42). Attempts to establish forts along the coast of the Gulf of Mexico proved to be futile because each time— in 1545, 1551, 1553, 1554 and 1559—storms destroyed the Spanish fleet. Unable to establish a beachhead on the Gulf Coast, the Spanish did manage to establish a fort at St. Augustine in what is now Florida. When the French attempted to take over the fort in 1565, they were stopped by a giant hurricane that destroyed their French fleet (42), leaving Spain to dominate the coast for many years.

Along the Atlantic coast there was a cluster of hurricanes from 1940 to 1966, averaging more than one every two years. From 1966 to 1997, only five major hurricanes struck the coast. This small number of hurricanes has corresponded with the enormous growth seen along the East Coast. The more recent incidence of damaging hurricanes has made the coastal population much more aware of the destruction hurricanes can bring.

Sea level rise and global warming

In recent times much concern has been expressed about the potential impact of sea level rise and global warming on coastal regions of the world. Although changes in sea level are known to have occurred cyclically over millions of years, events demonstrate that, since the furthermost equatorial advance of the Pleistocene ice masses some 16,000 to 18,000 years ago, global sea level along the East Coast of the United States was about 130 to

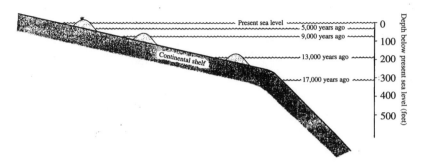

Figure 7.4. **The rise in sea level during the past 17,000 years** (7).

140 meters lower than at present (figure 4). The subsequent increase in sea level since then coincided with the polar retreat of the glacial ice masses due to melting, with the resultant water being poured into the ocean and raising sea level.

Various hypotheses have been put forward to explain these changes over geologic time. One theory predicts that ice sheets grow when the amount of sunshine reaching the north polar region is reduced, especially during the summer melting season, because of variation in Earth's orbit. The tilt angle of the earth's orbit varies between 22 and 24.5 degrees with a period of about 41,000 years, and the day of the year when the earth is closest to the sun (about January 3) varies on a 23,000–year cycle (43, 44). Currently interest in sea level change has been stimulated by concern over potential global warming. The earth's average surface temperature has risen approximately 0.6°C (1°F) in the last century, and the nine warmest years have all occurred since 1980. Earth's climate is controlled by incoming solar radiation and complex interactions of numerous factors. One factor determining climate is the presence of atmospheric gases that influence sunlight entering the earth's atmosphere. These gases also absorb some of the long-wavelength infrared energy reradiated from the earth's surface to outer space. Increased levels of gases, resulting from human activities such as the ever-increasing burning of fossil fuels, especially since the Industrial Revolution began in the 1750s, and deforestation activities, have trapped heat and warmed the earth's temperature. This phenomenon is commonly known as the *greenhouse effect*. Levels of carbon dioxide, one of the most important greenhouse gases, have significantly increased the past two hundred years and will continue to increase unless international agreements are approved and implemented to curtail its emission. Other gases contributing to the greenhouse effect are methane, nitrous oxides, and chlorofluorocarbons (CFCs), which result from various agricultural and industrial practices.

At the same time the earth's temperature is increasing, sea level along much of the United States coast has already risen 2.5 to 3.0 millimeters per year (10 to 12 inches per century). Scientists have predicted that:

- Global warming will most likely raise sea level 15 centimeters (5.9 inches) by the year 2050 and 34 cm (13.3 inches) by the year 2100 due to factors causing greenhouse warming;
- There is a one percent chance that global warming will raise sea level one meter in the next one hundred years and four meters in the next two hundred years;
- By the year 2100, climate change is likely to increase the rate of sea level rise by 4.2 mm/yr;
- Stabilizing emissions of greenhouse gases by the year 2025 could cut the rate of sea level rise in half; and,
- Factors other than anthropogenic climate change, including compaction and subsidence of land, groundwater depletion, and natural climate variations, will cause the sea to rise more than the rise resulting from climate change alone (45). Based on various earlier projections, if the Antarctic and Greenland ice sheets continue to melt, the oceans will flood many of the world's major cities (46). Figure 4 graphically depicts how this change in sea level might alter the coastal regions of the eastern United States. This potential impact has obvious implications to coastal zone resource managers in designing long-term development programs in both developed and developing countries (47).

In summary, some human interventions in the coastal zone have been utilized for many generations, but it is only within recent years that the potential environmental impact of these interventions has become a major concern. Two of the most common interventions over the years have been dredging and habitat modification. While the need for dredging for recreational and commercial activities is well recognized, there is also concern for the problems associated with them: increased amounts of silt clogging harbors and other waterways, the question of where to put the dredge spoil, and the loss of habitat for plants and animals.

As development activities increase in response to population growth, the pressure to build marinas and docks in coastal residential areas, many of which are vulnerable to the ravages of storms, becomes even greater. These construction activities do not occur without environmental cost. Both benthic communities and oyster beds can be substantially changed and/or destroyed. With construction of docks and marinas comes increased boat usage, which is usually accompanied by stirring up the sediments. In this manner pollutants, including anti-fouling paints and disposal of human waste from sanitary systems on boats, are introduced into the environment. The need to protect the environment associated with construction of marinas and docks has been recognized by the United States Environmental Protection Agency, and measures which would minimize negative impacts have been set forth.

Coastal beaches have long attracted human settlement. The desire to be at the beach has further intensified over the past few decades as population pressures have increased. Open beaches, however, are by their very nature dynamic and unstable habitats. They are built up by deposition of sand suspended in ocean water or eroded away when waves remove sand. In an effort to exert control over beaches, numerous efforts have been made to stabilize and/or to replenish eroding beaches.

Introduction of foreign and exotic species into a new marine environment often, but not always, has a detrimental effect on indigenous species. Some species are introduced deliberately, others accidentally, carried by ocean currents or associated with shipping activities. For an introduced species to successfully compete in a new environment, the invading species must be able to adapt functionally to the new ecological habitat conditions.

Often conflict arises between commercial fisheries and recreational fishers, especially in estuaries and nearby coastal waters. Unregulated fishing by each group can diminish or deplete important food species. In areas where aquaculture is an active commercial activity using non-indigenous species, care must be taken to prevent the release of these species and any accompanying pathogens into a new geographical region.

Only recently the value of coastal wetlands has been recognized and efforts made to restore at least some of them. The dominant species in southeastern salt marshes of the Atlantic and Gulf coasts of the United States is the smooth cordgrass, *Spartina alterniflora*, and progress has been made in successfully introducing this species in reclaimed salt marshes. The process of making the new assemblage of *Spartina* a "normal" functioning salt marsh ecosystem is, however, a long-term project, probably taking several decades.

Hurricanes and storms affect the eastern coastal regions of the United States and the Gulf Coast and have been known to do so since the time of Columbus. The frequency of hurricanes varies. From 1940 to 1966 hurricanes occurred more than once every two years on average, while only five major hurricanes struck the coast from 1966 to 1997. The smaller number of hurricanes has corresponded to the time of most explosive population growth along the coast. By contrast, storms, especially "nor'easters," occur yearly and greatly influence sand transport resulting in beach erosion. Developed areas—housing and commercial sites—are much more likely to be affected than undeveloped barrier islands and marshes.

The potential for global warming has raised concern about the impact of sea-level rise on coastal regions. The earth's average temperature has risen approximately 0.6°C (1°F) in the last century, and the nine warmest years have occurred since 1980. As the earth's temperature has increased, sea level along much of the United States coast also has risen from 2.5 to 3.0 mm per year, or 10 to 11 inches per century. The warming of the earth has been linked to the burning of fossil fuels and other gases produced by industry and agriculture and to deforestation activities. This combination of factors trap the heat and thus contribute to warmer temperatures and rising sea level.

8

Policy and Socioeconomic Implications Affecting the Coastal Zone

Natural processes may shape the geomorphology of the coast, but environmental policies determine the look of the landscape. Both the public and private sectors influence formulation of coastal environmental policies, especially in a nation with as much coastline as the United States.

In the public arena, the responsibility for managing the coastal zone falls to all three jurisdictional levels of government. Governing jurisdictions espouse policy to set forth their beliefs concerning issues that face them. Environmental management of natural resources is just one of the areas governing bodies must address in declaring the way business will be conducted within their jurisdictions. Not only is the management responsibility spread over different levels of government, but different agencies at each level play a role in the process. The private sector plays a significant role in the political process by exerting influence as statutes and regulations are considered by the Congress, state legislatures, and local governing bodies.

Environmental policy for the coast is espoused through these statutes, regulations, and ordinances. Effectiveness of these policies depends on their energetic administration by those charged with implementation. A watchful public is the only assurance that environmental policy is implemented as envisioned by the governing statutes.

The federal role
The National Environmental Policy Act

One of the first modern-era attempts by Congress to recognize the federal responsibility in protecting the environment was codified as the National Environmental Policy Act of 1969 (NEPA). The statute sought "to create and maintain conditions under which man and nature can exist in productive harmony, and fulfill the social, economic, and other requirements of present and future generations." Thus, national attention was focused on environmental quality, and the federal government was put on notice that every action contemplated must consider its impacts on the environment. The act established the council on environmental quality within the executive office of the president. The council is charged with assisting the president in preparation of the annual environmental quality report, which reports on the state and condition of the environment. The statute galvanized the federal government's interests in environmental matters. The law requires federal agencies to be cognizant of NEPA and adhere to its requirements in all their undertakings. Most of the important environmental legislation in this country has been enacted since NEPA became law in 1969. With a bureaucracy as large as that of the federal government, it is not surprising to find programs that work at cross purposes. Such is the case when one examines the statutes, regulations, and programs that impact coastal environmental policy.

In 1979, President Carter directed the National Oceanic and Atmospheric Administration within the Department of Commerce to conduct a systematic examination of federal programs which have an impact on the coastal zone of the United States. NOAA issued the results of the study as the *Federal Coastal Programs Review*. The report identified a myriad of federal programs influencing the coastal environment. Some of the programs assisted with conserving natural resources, while others subsidized development of some of those same resources. The results of the exercise led to a reexamination of the role the federal government plays in shaping the future of America's coasts. Although some programs were modified to reduce their subsidies to development, federal appropriations still play a major role in determining the look of the coast.

A review of federal laws directly affecting the coastal zone yields a list of surprising length. The following laws and programs are generally considered most important.

The Coastal Zone Management Act

An outgrowth of the 1969 Stratton Commission's report, *Our Nation and the Sea*, The Federal Coastal Zone Management Act of 1972, as amended, (CZMA) is really the only true federal land-use program affecting private property. It is ironic that this statute was enacted at about the same time land use bills sponsored by Congressman Morris Udall and others failed miserably. The impetus provided by the Stratton Commission's work cannot be underestimated. *Our Nation and the Sea* focused the attention of scientists, citizens, and political leaders on the importance of the nation's coasts and the improved management required. The CZMA establishes a voluntary program with federal assistance to states that desire to develop management programs for their coastal zones. The most rigorous requirement a state must meet is that its program strikes a balance between development and the environment. Methods to achieve this ultimate goal are left to the individual states. The framers of the act argued that the coasts of no two states are exactly alike; therefore, the management plans should not have to be exactly alike, either.

As incentives to make participation in the federal program attractive, states are provided grants-in-aid, on a matching basis, to defray the cost of program development and implementation. Over the years, the match requirement for participating states has increased from one-third to one-half the cost of the program. Yet the program has been quite successful. As of 1998, thirty-one of thirty-five eligible states were participating in this collaborative effort to manage this country's coast.

Another strong incentive for state participation is Section 307 of the CZMA. It requires federal actions in the coastal zone to be consistent, to the maximum extent practicable, with a state coastal management program subject to approval by the secretary of commerce. Thus, a state program must contain enforceable policies that grant a certain degree of predictability to the entire management process. Citizens, as well as federal agencies, know what to expect when proposing activities in the coastal zone. For federal agencies, consistency applies to direct federal activities like Corps of Engineers navigation and flood control projects. It also applies to permits and licenses issued by federal agencies. Federal financial assistance and activities on the outer continental shelf are also subject to the Section 307 provisions of the CZMA. Several federal consistency cases have won notoriety. Most of these have involved continental shelf oil and gas development or dredged spoil material disposal.

Despite the controversy the CZMA consistency provisions have engendered, they remain a powerful tool for management of the coastal zone. Disagreements arise mostly from a reluctance on the part of federal agencies to submit to state officials and their views on management of resources. Federal agencies are quick to invoke the tenants of "national interest" and "national security" in arguing with the states. The 1990 amendments to the CZMA attempted to clarify the role of federal consistency in management of the coast.

The CZMA has undergone several changes through the reauthorization process since its enactment in 1972. The 1976 amendments added three planning requirements for state programs: states were required to consider shoreline access, shoreline erosion, and energy facility siting in developing their programs. Additionally, Congress added Section 308, the Coastal Energy Impact Program (CEIP), to assist states in coping with expanded oil and gas production impacting the coastal zone. Project Independence, a plan to expand domestic petroleum production, spurred exploration on the energy-rich outer continental shelf. Ultimately, this petroleum traversed the coastal zone to refineries. The CEIP helps ameliorate the social, economic, and environmental impacts in affected coastal states.

In 1990, Congress added Section 6217 to the statute in an attempt to address coastal non–point-source pollution. The initiative differed from the existing provision codified as Section 319 of the Federal Clean Water Act. The Coastal Nonpoint Pollution Control Program required enforceable policies to address the problem. Efforts to implement this provision have been frustrated by opposition from some states and interest groups. These states prefer a more voluntary approach to non-point pollution control. Additionally, state water quality agencies are turf-conscious when they feel CZM programs are becoming too involved in the non-point pollution control business. The joint administration of the program by Environmental Protection Agency and NOAA also caused squabbles. The opposition has successfully diluted the enforceable nature of the program with amendments in the latest reauthorization of the Clean Water Act. Because of these difficulties, effectiveness of the program has yet to be finally determined.

A strong statement of national policy concerning America's coasts was contained in an earlier reauthorization of the statute. The Congress enumerated the following areas as items in the national interest:

- natural resource protection
- hazards management
- major facility siting
- public access
- urban waterfront and port redevelopment
- simplification of decision procedures
- intergovernmental coordination
- public participation
- living marine resource conservation

This was a significant statement in that it helped guide the implementation of CZM programs by participating states until the act was reauthorized in 1990, when Congress identified the following priority management areas:

- increasing opportunities for public access
- planning for ocean resources
- protection/creation of new coastal wetlands
- mitigation of natural hazards
- reducing marine debris
- preparation/implementation of special area management plans

• addressing secondary and cumulative impacts on coastal resources
• siting energy facilities

Federal appropriations for implementation of CZM, because of strong congressional support, have enhanced the program. States that have benefitted most from the program are those that have exceeded the 50 percent match requirement. These are usually the stronger, more influential programs with regard to coastal resources management. NOAA has attempted to reward these states with strong programs through the Coastal Zone Enhancement Grants Program.

The CZMA continues to be the cornerstone upon which state CZM programs are built. The law has evolved through several reauthorizations since 1972. Congressional support, even by some very conservative Republicans, portends well for a continued federal role in CZM through this statute.

U.S. Army Corps of Engineers' programs

While the CZMA represents a landmark in coastal resources management, several other federal statutes also have significant policy implications. One of the earliest programs affecting coastal resources is the Corps of Engineers' responsibility to protect and maintain the navigability of the waters of the United States. This responsibility is derived from the commerce clause of the U.S. Constitution, which empowers Congress to regulate commerce. Courts have held that the federal government has the right to regulate navigation because it is so important in promoting commerce. The Corps of Engineers is the federal agency charged with the responsibility of enforcing the government's right of navigation servitude. The Corps accomplishes this through civil works projects and regulating activities in navigable waters through a permit system. The Corps relies on the Rivers and Harbors Act of 1899 as its legal authority to carry out this responsibility. In the 1970s, the Corps was charged with additional responsibilities under Section 404 of the Clean Water Act, significantly expanding the federal role in management of navigable waters to include wetlands adjacent to navigable waters.

Another area of Corps responsibility that has policy implications for the coastal zone is the shoreline protection program. Projects range from construction of hard structures like seawalls to soft solutions like beach renourishment. State governments have also been active in providing funds to construct these projects. These federally subsidized activities tend to give a false sense of security to coastal property owners. The soft solutions are usually short-lived, and the long-term erosion of beaches remains a problem. Substantial sums of money are expended annually on such projects, with the major justification that they represent capital investment in the tourism business and local economies. Coastal property owners and oceanfront local governments have been quite successful in convincing government leaders that this is a wise investment. It represents a federal subsidy with policy implications with regard to siting of structures along the coast—especially those areas that experience chronic shoreline erosion.

The Clean Water Act

A true milestone in environmental regulation was reached with the passage of the Federal Clean Water Act in 1972. This broad-ranging statute addresses such issues as point-source pollution control, ocean discharges of effluent, non-point pollution control, water quality standards, and dredge-and-fill operations associated with navigable waters. The statute has significant policy implications for the coastal zone. Of particular importance is the regulatory program established for wetlands. The Corps of Engineers administers this Section 404 program, but EPA and the U.S. Fish and Wildlife Service are important players in this regulatory scheme. EPA actually has the authority to remove a wetland area from the Corps' jurisdiction should the agency feel so strongly that a permit for a proposed dredge and fill operation should not be granted.

The anti-degradation clause of this law also has policy implications for the coastal zone. This provision requires the government to protect existing uses supported by a water body. Marina development in waters supporting shellfish propagation was significantly impacted in the 1980s by this provision. The federal Food and Drug Administration requirement to impose a buffer area around marinas for public health purposes influenced the decision by EPA to oppose marina construction in shellfish waters. FDA, concerned with public health, feared contamination of the shellfish by fecal coliform bacteria from waste facilities aboard vessels using the marina. The agency determined that an existing use, such as shellfish propagation, was protected by the water quality classification system. To allow construction of a facility in those waters that required a closure area where no harvesting could legally occur was a violation of the anti-degradation clause, it concluded. Thus, a permit to allow marina construction could not be issued by the Corps because the action would violate this important provision of the statute. As a matter of policy, EPA had significantly affected siting of marinas in the coastal zone. Resort developments prefer pristine waterfront locations— but these same pristine waters are likely to be classified as shellfish harvesting waters. Thus, the conflict arises when marinas as amenities for resort developments cannot be constructed in these waters. Here, a resource protection issue greatly influenced the development pattern of boat-docking facilities at resorts especially in the Southeast.

The Clean Air Act

The Clean Air Act has undergone significant changes since its enactment in 1972. The statute has policy implications for the coastal zone because its requirements often determine development and growth patterns. Of particular importance is a policy espoused by Congress "to preserve, protect, and enhance the air quality in national parks, national wilderness areas, national monuments, national seashores, and other areas of special national or regional natural, recreational, scenic, or historic value" (2). Since many of these protected areas are located within the nation's coastal zone, the policy impacts coastal development patterns. Activities such as industrial development, which may cause air emissions, are impacted if those emissions cause degradation of air quality in one of the protected areas.

The oil and gas industry is affected particularly in states like California where some areas that have not attained air quality goals prescribed under the statute. This statute also affects oil and gas development on the outer continental shelf. In order to site certain facilities, such as a refinery, additional potential air emissions from the proposed facility may cause the air quality standards for the area to be exceeded. Again, this is an example of how coastal development patterns may be affected by the policies of the Clean Air Act.

National Flood Insurance Act

This program, administered by the Federal Emergency Management Agency's Federal Insurance Administration, affords subsidized insurance against damage by coastal flooding to owners of real property. Ironically, some of the wealthiest American owners of expensive coastal properties receive this subsidy. In many ways, the program functions as an incentive for development in hazard-prone coastal areas. The federal rationale behind the program is to provide subsidized flood insurance if communities will adopt enforceable policies to address minimum floodplain management standards. These standards include restrictions on construction in floodplains and requirements for elevation of structures above a base flood elevation to reduce potential damage from flooding. Structures in the floodplain that are financed through federally insured and regulated financial institutions are required to carry this insurance. Similarly, properties with federally guaranteed mortgages are mandated to have this coverage.

One of the policy shortcomings of the program in coastal areas has been the failure to consider long-term shoreline erosion. Efforts to include this facet of risk in the program have been frustrated by special-interest groups who do not want the issue addressed. Proponents argue that erosion risk should be considered as much a risk along shorelines as flooding. Afraid of premium increases, property owners in several locations have successfully organized politically to prevent Congress from mandating that the Federal Insurance Administration include property damage from erosion as a major component of the program.

Flood insurance can be obtained from private sources, but at a much higher cost. Obtaining the coverage is sometimes difficult, as well. Federally subsidized flood insurance has had a profound effect on coastal development. The federal government has, as a matter of policy, assisted the development of America's coasts by assisting in insuring developers against their major risk—damage from coastal flooding.

Federal Disaster Assistance

Another federal initiative that has done nothing to discourage development in coastal high hazard areas is the longstanding commitment to assist state and local governments in coping with national disasters. The Federal Emergency Management Agency (FEMA) has the overall responsibility for providing relief to individuals and to public entities to assist with recovery from disasters. The major public policy issue is the federal government's assuming of a portion (75 percent in most cases) of the risk associated with public investments in the coastal zone. Examples of public investments eligible for reimbursement in case of disaster include publicly renourished beaches, water and sewer systems, roads, bridges, recreational facilities, and buildings like libraries and hospitals. Communities must be participating in the National Flood Insurance Program to be eligible for reimbursement.

Disaster assistance has policy implications with regard to the initial siting of public facilities. The knowledge that the federal government will reimburse communities for losses does little to encourage siting of facilities elsewhere, outside coastal high-hazard areas. The ability of a community to pay is not a consideration in the process, and politics usually dictates an early disaster declaration by the president when a natural catastrophe does occur. Several other federal agencies, such as the Small Business Administration and the Department of Education also provide assistance.

To the credit of FEMA, the agency has attempted to reform the program in recent years. Under the latest reauthorization of the governing statute, Congress has placed more emphasis on mitigation of hazards. Hopefully, this approach will eventually lead to reduced assistance outlays as a result of national disasters.

The National Estuarine Research Reserve Program

Section 315 of the CZMA established a national system of field laboratories for the purpose of studying coastal ecosystems. The goal of the initiative was to gain information on the natural processes occurring in America's estuaries. Additionally, the effects of human activities on these environments could be assessed if a baseline for pristine estuarine conditions could be established. Public education was also a major reason Congress established this network of reserves. The system has policy implications, since properties surrounding these reserves are impacted by their proximity to them. Reserves are usually designated as "geographic areas of particular concern" under the CZMA by states, thus affording them more regulatory protection.

The National Marine Sanctuaries Program

Created in 1972 by the Marine Protection, Reserve and Sanctuary Act, the National Marine Sanctuary Program is designed to afford protection to nationally significant marine areas. In order to be considered for inclusion in the system, an area must possess unique historical, educational, research, ecological, recreational, or aesthetic value. To make new sanctuary designation more politically palatable, historical uses within the sanctuaries, such as recreational boating, are usually protected. Uses are either allowed throughout the sanctuary or prohibited. This multiple-use concept has caused heated debate during the nomination process for new sanctuaries. Some legislatures have attempted to use the program to exclude oil and gas development on the outer continental shelf. The program has policy implications due to its goal of establishing restricted marine areas to protect unique qualities. The program's objective of fostering public education about marine resources is being realized. Although hampered by lack of adequate funding, the program enjoys strong congressional support.

The Coastal Barrier Resources Act

The Federal Coastal Barrier Resources Act was an outgrowth of President Carter's Coastal Programs Review. The Department of the Interior was charged with administering this program which reduced federal subsidies for development of undeveloped barrier islands and barrier beaches. Congressional policy intended to reduce the development pressure on undeveloped barrier islands. The aim was to reduce storm-related damage to people and property, resulting in a reduction of public expenditures to cope with the aftermath of such disasters. Another objective was to preserve significant wildlife habitat harbored by these areas. The Coastal Programs Review identified several federal programs that aided and abetted development of these barriers: low-interest-rate loans from agencies like the Farmers Home Administration and flood insurance were identified as just two of the many culprits. The Coastal Barrier Resources Act charged the Department of the Interior with mapping the coastal barriers and submitting the system for congressional approval. Today there are more than five hundred units in the system. Federal expenditures and subsidies are severely restricted in these areas.

Several studies have evaluated the effectiveness of the statute in curtailing development on these barrier islands. In general, results indicate the program has been quite effective in reducing federal incentives to development. However, several of the units in the system have been developed without the federal subsidies to include flood insurance. This statute did mark the most significant attempt by Congress to limit the federal government's subsidizing often inappropriate development in the coastal zone.

The National Estuary Program

This program was included as part of the reauthorization of the Clean Water Act in 1987. An outgrowth of earlier EPA work in the Great Lakes and the Chesapeake Bay, it is an attempt by the Federal government to address water pollution problems on a more regional basis. Federal funds are made available for development of a management plan for the area and states are responsible for the implementation of the plan. An elaborate system of committees is established to address the problems encountered in a given study area. The Comprehensive Conservation and Management Plan is the ultimate product of this process. Unfortunately, the jury is still out as to effective implementation of the plan. The program seems to have been more successful at identifying water–pollution related problems than solving them. The difficulty arises in convincing states to adopt enforceable policies to address these problems.

CZMA Special Area Management Plans

The development of Special Area Management Plans (SAMP) and their subsequent implementation has received mixed reviews. The primary objective of this program is to afford a mechanism to address coastal environmental problems that transcend jurisdictional boundaries. A good example of this process in action is the Charleston Harbor, South Carolina Special Area Management Plan. This plan addresses specific management goals in an area encompassing multiple jurisdictions. Federal, state, and local governments each have a stake in Charleston Harbor, and each is represented on the management committee that oversees the project. The process differs from most environmental programs in that it examines the social and political as well as the natural systems within the study area. The primary success of an SAMP lies in the ability of all jurisdictions to agree on the problems facing them and to work collectively to address them.

The Endangered Species Act

One of the most controversial environmental laws to be enacted is the Endangered Species Act. Initially enacted in 1973, the statute has undergone intense scrutiny by the Republican controlled Congress in recent years. Its primary goal is to conserve endangered and threatened species. The law has tremendous policy implications for the coastal zone in that land uses are affected in carrying out its mandates. The statute was highlighted when loggers in the Northwest were precluded from harvesting certain timberlands that represented critical habitat for the Spotted Owl, an endangered species. Real estate developers in the Southeast know the impact the presence of the Red Cockaded Woodpecker has on development plans. The law continues to be controversial.

Other federal policy instruments

Several other Federal programs have significant policy implications for the coast, though some of their impact is sometimes slight. Among these are subsidized loans from the Department of Housing and Urban Development and the Farmers Home Administration. EPA provides grants for water and sewer construction. The Department of Transportation provides grants for highway and bridge construction.

The federal tax code is another source of subsidies for coastal development. Property taxes and interest are allowable deductions that increase the attractiveness of investment in second homes. The casualty loss deduction also shifts some of the risk for building in the coastal zone to the taxpayer. Real property owners are allowed to deduct a loss incurred in a storm minus the proceeds of any insurance. There are some limitations on the amount of the deduction, but it remains a subtle subsidy to coastal growth even in high-hazard areas.

The state role

Regulation of land use and natural resources management has largely been the domain of state and local governments. This fact was recognized and respected by the CZMA. Although the federal government plays a significant role in policy formulation, the states and local governments have been made major players in the coastal management process by enactment of the CZMA.

Through the CZMA, states are encouraged, and supported financially, to be more aggressive in management of coastal resources. The degree of effectiveness of state coastal management programs varies, as one would expect, but many states have become more effective players in the management game.

Most states have aggressively pursued wetlands protection. Many have enacted laws even more restrictive than the Section 404 program administered by the Corps of Engineers. Several states have refused to certify Corps nationwide permits that make it easier for applicants to alter wetlands as consistent with their CZM plans. States are becoming leaders in efforts to require mitigation for impacts to wetlands and in establishing wetland mitigation banks to facilitate implementation of this concept.

Shoreline management is one area of endeavor where states have excelled. States like Michigan, North Carolina, Florida, and South Carolina have led the effort in informing the public of the hazards associated with building along the shore. The important tourism industry fostered largely by access to beaches has spurred this effort. States realize the importance of the dry sand beach to the tourism industry. Many states now require construction setbacks from the beach. States like South Carolina have adopted retreat policies requiring property owners to move landward after natural disasters destroy structures. These policies have not gone unchallenged, however. The Supreme Court's 1992 decision in *Lucas v. South Carolina Coastal Council* dramatically impacted the course of the regulatory process and gave property rights advocates new ammunition. Nonetheless, state efforts in managing the shoreline have been largely successful, and as states refine their programs, the process should become more sophisticated and should yield more defensible methodologies to determine allowable rebuilding of structures after natural disasters. State policies that encourage renourishment of beaches, where feasible, have proven an effective adjunct to the harsh regulatory requirements of retreat.

Provision of beach access to the shoreline for the public is another area where states have been effective. States like California and Texas have been especially aggressive in this area of coastal management. The California Coastal Commission made history through its involvement in a 1987 Supreme Court case, *Nollan v. California Coastal Commission*. The commission sought to condition a seawall permit to require a lateral access easement along the beach in front of the proposed wall. The effort was rebuffed by the Supreme Court which found that the imposition of such a condition constituted a taking of the Nollans' property without just compensation. Coastal regulators now realize any condition imposed on a permit must relate directly to the impact of the development.

Other states have undertaken aggressive land acquisition programs. Florida has expended millions of dollars to acquire beaches to enhance public access. California, through the efforts of the State Coastal Conservancy, has been a leader in utilizing innovative techniques to afford improved public access to coastal resources.

Oregonians have distinguished themselves by assuming the leadership role among the states in ocean management. Other states, like North Carolina, are attempting to become involved in this area of coastal management. States are realizing that activities off their shores have the potential to impact valuable coastal resources. Looking offshore, states see important marine habitats, marine mammals, and potential oil and gas reserves. These are some of the reasons states are beginning to pursue the expansion of their management purview into the adjacent ocean.

States are becoming increasingly active in formulating coastal environmental policy. They realize that they possess a more detailed knowledge of the state political climate than do federal agencies. Crafting laws and regulations that take into account differences among states makes sense. In reality, even though more aggressive in formulating enforceable policies, states still rely on federal funds and federal agency cooperation to insure enforcement of their policies.

The local role

Local governments have historically shouldered the primary responsibility for determining land use and managing natural resources within their jurisdictions. Local governments have carte blanche authority over land and water uses within their boundaries. They make most of the decisions that directly impact the average citizen.

Increasingly, citizens are realizing that public involvement in local government is mandatory if they want to influence the future of their communities. Municipalities are springing up throughout America's coastal zone for that exact reason. Citizens want decisions concerning their communities made as close to the electorate as possible. Thus small municipalities afford an opportunity for community involvement in determining the future. On the South Carolina coast, for instance, new municipalities have been formed in recent years at Litchfield Beach, Pawleys Island, Hilton Head Island, Kiawah Island, and Seabrook Island. These new towns were created by residents who wanted a greater say in establishing environment and growth policies for their communities. They were not satisfied by the controls on development afforded by their respective county governments.

Traditional zoning and subdivision regulations have long formed the basis for most local control of growth. Local governments are increasingly turning to innovative methods to regulate development. These include the institution of urban growth boundaries and planned unit developments. Some coastal communities are using incentive and performance zoning as well as developer agreements in an effort to work with the development community to attain mutual goals. Communities like the town of Hilton Head Island have undertaken aggressive land acquisition programs. Hilton Head used revenue from a real estate transfer fee to buy up land, thus precluding its development. Not only does this process reduce the total number of dwelling units on the island, it reduces traffic congestion, a major problem for this municipality. Additionally, it provides open space for island residents and visitors. Other local governments are acquiring less than fee-simple interests in property to regulate growth.

The imposition of impact fees by local governments is a relatively new policy initiative to help control growth. Fees charged for new development help defray the cost of infrastructure like water, sewer, and roads. Since installation of infrastructure is probably the primary determinant of growth in the coastal zone, local policies about location and timing of infrastructure will increasingly determine how the coast is developed. Density of development may be a more important factor in determining environmental quality than new development itself. In controlling infrastructure, local governments often make the seminal decisions determining the ultimate quality of the coast's environment.

Increasingly, policies established by local governments are becoming the greatest determinant in shaping the future of this nation's coasts. Public involvement in this process is imperative if citizens expect to have a voice in their community's future. It is far easier to talk to one's town or county council member than to influence a member of Congress. Therefore, the greatest determinant of environmental quality should be a body making policy at the level closest to its citizens. Local governments wear this mantle. It is incumbent upon local residents to become involved in policy formulation if they wish to influence the future environment of their communities.

Role of the private sector

Environmental policy formulation is dramatically influenced by the private sector of the U.S. economy. Most major corporations have full-time environmental and governmental relations staffs. These individuals monitor proposed actions by all levels of government and utilize the political process to dispose of these proposals in a fashion favorable to their employers. The private sector, as is their right, has become very adept at cultivating political leaders who will lend a favorable ear to their point of view. Campaign contributions form the basis for many of these relationships.

Non-governmental organizations (NGOs) also constitute a powerful force in environmental policy development. Groups like the Natural Resources Defense Council, the Sierra Club, the Conservation Foundation, the Nature Conservancy, the National Wildlife Federation, and the League of Women Voters wield great influence with many legislators. These special-interest groups, just as business-oriented interests, are well organized and in many cases well financed. In recent years conservation groups have begun to grade legislators on each conservation vote. Pro-business groups have also employed this tactic to hold elected officials accountable for their votes.

NGOs seem to be gaining in influence in the policy-making arena. They have become adept at working with agency personnel as well as elected officials in manipulating the process to gain their desired results in environmental policy.

Socioeconomic considerations

The coastal zone of the United States is experiencing unprecedented growth. Estimates place the population of the narrow band where land and water interact to include almost 70 percent of the country's total early in the twenty-first century. With this development come environmental challenges for all levels of government. Of particular concern is the social and economic impact attendant to this growth.

Many of the new inhabitants of the coastal zone will be retirees. This group brings with them unprecedented wealth accumulated in retirement plans that have benefitted by successful investments in a strong economy. Retirees also bring with them high demands for local services. The aging population will strain the health care system of many states. The years ahead will yield coastal populations with the highest percentage of retirees ever experienced. Many of these retired men and women formerly held responsible positions in business, industry, government, and the military during their working years. Community leaders should seek to tap their intelligence and energy for social and political good.

Of particular concern to some sociologists is the scarcity of good-paying jobs in the resort-oriented economies that have been developing along portions of the coast. The higher the percentage of retirees in a coastal population, the more likely it is that opposition will crystallize against proposals for industrial development in the community. The "blow the bridge" philosophy pervades some communities where residents have found the good life and desire to freeze their experience in time. Economies based largely on tourism have experienced difficulty in creating high-paying, permanent jobs. While significant portions of the coast sustain traditional uses that are water-dependent, the new era of growth in resorts has supplanted these in certain instances. Shellfish harvesters find themselves out of work when coastal urbanization contributes to closure of shellfish grounds. Shrimp boats find they are not welcome near certain resort beaches. Trays placed in tidal creeks holding species propagated as part of the mariculture industry may have a deleterious effect of the aesthetics of certain resort areas.

The environmental impact of coastal growth should be addressed by serious, deliberate formulation of rational policy. Confronting the social and economic aspects of this development requires no less study and public involvement.

Coastal environmental policy, then, is the product of a process that allows divergent views to be aired and considered by elected officials and regulatory agency personnel. The process exists at the federal, state, and local levels of government. The private sector, including NGOs, holds great influence in this policy-making process.

Formulation of environmental policy at three levels makes it almost inevitable that some programs and initiatives will work at cross purposes. An informed public that vigilantly scrutinizes the implementation and administration of environmental laws is the nation's best insurance against ineffective environmental protection. Every citizen bears this responsibility and should take it seriously. The democratic freedom to participate in the policy formulation process has been won by the sacrifices of many. American citizens who care about the coast should be mindful of this privilege and avail themselves of this opportunity.

To provide the reader with a more detailed description of the complexity of coastal zone management problems and procedures, selected references are cited (3 to 12).

9

The Future of the Coastal Zone

Forever changing, both over geologic time scales due to natural causes and over shorter time as a result of human intervention, the coastal zone has adapted and persisted in its present day form. But what is the future of the coastal zone?. Will the impact of the crush of increased human population in the coastal zone continue on a piecemeal basis with little planning, or will human society realize the virtue and necessity of resource management and planning before environmental disasters occur? Will scientific findings become an important additional factor in influencing resource planning, or will emotional and political reasons prevail? What will be the short- and long-term costs associated with providing the necessary infrastructure to support the increase in population?

The major force determining the utilization of coastal resources is the continuing impact of increased human population. Every forecast predicts significant movement of people to coastal zones. For example, an increase of as much as 48 percent is forecast for the coastal region of South Carolina by the year 2010. Similar predictions have been made for other coastal regions of the United States and the world. With this increase in population density, the demands and competition for finite coastal resources will continue to escalate. In 1995, NOAA estimated that by the year 2000 the following changes could be expected in the coastal areas of the United States:

- The demand for potable water will increase by at least 10 percent;
- The demand for energy will increase by at least 12 percent;
- Sewage will increase by 18 percent;
- Seven million more houses will be constructed; and
- Eleven million more cars will be on coastal highways (1).

It is clear that continued population growth will exacerbate existing environmental problems and undoubtedly create new ones. Many problems associated with growth and development, particularly in the southeastern United States, are already occurring: harmful algal blooms (HABs); hypoxia events; loss of wildlife habitat and economic resources; flooding and storm disasters; loss of historic places and cultural context; and enormous financial costs to abate beach erosion, flooding, and contamination. The frequency and intensity of such catastrophic events are increasing as development of the coastal zone proceeds. How society will deal with these problems is the fundamental issue. At present, examples of the negative results of unbridled development of coastal regions can be seen in many areas. These portend similar results for other coastal regions in the future: poor water and air quality, inadequate sewer and water treatment facilities, closed shellfish beds, clogged highways, and excessive habitat alteration. By contrast, examples of undeveloped pristine coastal regions still exist, and these permit society to see how such environments looked in the past and provide a sites for studying how undeveloped coastal systems function. This information is vital for such activities as bioremediation, habitat restoration, and development of predictive ecosystem models.

A new societal ethic must emerge that is founded on the premise that utilization of finite resources will be based on planning that incorporates scientific, socioeconomic, and cultural data and values. An essential requirement of this ethic is that citizens take an active role in determining the destiny of the coastal zone. This book stresses the scientific bases which hopefully will permit an informed citizenry to better determine the future of the coastal environment.

Future continuing problems

Although population growth is the basic problem confronting the health and integrity of the coastal zone, in 1994 the National Research Council (2) identified a number of current and persistent problems which need to be addressed even if the coastal population remains static:

- Eutrophication
- Habitat modification
- Hydrologic and hydrodynamic disruption
- Exploitation of resources
- Toxic effects
- Introduction of nonindigenous species
- Global climate change and variability
- Shoreline erosion and hazardous storms
- Pathogens and toxins affecting human health

To this list can be added other problems that have been identified by other groups concerned with coastal issues:

- Decreased availability of freshwater
- Failure to preserve existing natural environments
- Increased non-point pollution

- Proliferation of tourist related industries
- Increased unmanaged stormwater runoff

As discussed in earlier chapters, much research has already been devoted to these problems, but solutions to each problem still require further fundamental research in order to develop better regulatory and management procedures. One emerging concern that is receiving increasing attention pertains to assessment of multiple environmental impacts at the coastal ecosystem level rather than considering the effect of only one type of perturbation at a time. Some of the newer techniques of analyzing large datasets and developing systems models, such as those associated with Geographic Information Systems (GIS) and Geographic Information Processing (GIP), are proving invaluable (3, 4). Improved methods of using biological measures in assessing the impact of toxicants are being developed (5). With time, even more sophisticated techniques will evolve to facilitate integrative analyses.

Because of the difficulty in assessing whether changes in the environment are due to human interventions or long-term "natural" environmental phenomena, it is necessary to expand existing long-term environmental programs in both relatively pristine environments and those which are subjected to human-oriented perturbation. The responses of the pristine environment serve as an indicator of "natural" changes, and the responses of perturbed systems indicate human impacts. Well-designed long-term monitoring programs are also essential in detecting subtle short-term environmental changes that could have a cumulative long-term negative impact on ecosystem health. Short-term subtle changes can serve as early indicators of long-term system change. New techniques are being developed and deployed that can remotely record multiple environmental factors and transmit the data to laboratories for analyses. These techniques are not labor-intensive and, therefore, have generally proved more cost-effective than some current methods of sampling. These are a few examples of efforts to provide the scientific input needed for managing the coastal environment.

Solutions

The conflict between increasing population growth and development and environmental protection is rapidly becoming acute. Scientists have written extensively about the importance of protecting coastal areas, but simply making information available to the public does not ensure public action. It is true that there are laws in effect that do offer guidelines on both business and residential development, but many laws and regulations need to be regularly reviewed and revised to reflect scientific advances. Laws will not be effective if the will does not exist on the part of the public, the policy-management community, and the development industry (homebuilders, real estate agents, developers). Often the objectives of economically-based development projects and those of conservation efforts are very difficult to reconcile. The amount of coastal area devoted to residential, recreational, industrial, and conservation use is subject to constant and intense societal and economic pressures. How, then, can problems related to human access to coastal areas be resolved? One method which has had some success is collaborative planning (6).

Collaborative planning must involve active participation by those individuals representing all potentially competing interests—environmentalists, scientists, developers, regulators, and property owners, as well as others with a vested interest. Although a time-consuming and often an uncertain process, collaborative planning has been successful in resolving environmental and development issues. In addition to including all affected segments of society, such a planning effort also must have strong political backing and leadership. There must be continuity of planning and management as well as compromise—consideration of safety valves such as finding room for some development. In the Chesapeake Bay area, for example, the focus was on natural resource areas, but further development was directed toward already-developed areas (6). Funding for a collaborative planning efforts must also be considered. In addition to finding the money to initiate the process, there are data to be collected and technical reports to be compiled before making management decisions.

Some issues surface in any planning activity. Preservation of natural resource areas is contentious, since over 70 percent of U.S. wetlands are privately owned and the question of funding for acquisition and preservation inevitably arises (6). The purchase of open spaces and greenbelts is often deemed important in developing areas for aesthetic reasons. To the surprise of many, conservation of such areas has been found to increase adjacent residential property values (7). Some local governments finance land acquisitions through property and sales taxes, fees, service charges, and funds from special assessments. In some cases, local governments are able to borrow money, but this avenue is subject to voter approval which is not always forthcoming (7). Other options are to solicit grants from federal and various non-governmental agencies. Fodor (8) suggests that purchasing private lands and doing nothing with it is often less expensive for a community that providing the infrastructure for development—that is, development is extremely expensive for taxpayers.

Often communities welcome new developments with open arms. Then they find the infrastructure needs of rapidly developing areas often outstrip property tax revenues. As a result, communities must raise taxes for all residents. Another potential source of revenue to offset the costs of construction of roads and water and sewer lines are assessments of impact fees or exactions levied against developers. This is a controversial route; many argue that these fees are a burden to homeowners, who pay both the impact fees and property taxes designated for infrastructure.

Another essential activity of future planning for management of the coastal zone is to develop a holistic viewpoint of the environment: the coastal zone does not exist as an isolated portion of the earth's ecosystem but can be impacted by adjacent environments. However, coastal regions and the river basins, whose waters drain into the coastal estuaries and bays, frequently have management plans that have been developed independently of one another. As illustrated in chapter 7, some management practices used in the upper Mississippi River basin have harmful affects on the fishery industry in the Gulf of Mexico. River basins and coasts are linked through numerous natural and socioeconomic processes, such as the water cycle, sediment and chemical transport, and human activities. Integrated management of these two systems could result in better coordination of policy-making across sectors, for instance water, forestry, agriculture, urban development, and environmental protection (9, 10).

To correct previous practices that have resulted in destruction of critical coastal habitats, initial research efforts in the topics of bioremediation and habitat restoration need to be actively supported by various private and public granting sources. The important economic-environmental role of these habitats—as well as their destruction—has been well documented. Now it is important to improve the quality of the environment.

An essential aspect in providing a solution to future coastal issues is developing new education programs and expanding existing ones. Since the late 1960s and early 1970s, awareness of environmental issues has significantly and education programs have been developed in response to environmental concerns. These programs have reached all age groups and all levels of the formal and informal educational experience. As a result, the general public is more aware of and informed about environmental issues than ever before. However, there is a constant, continuing need to provide the public with new information and concepts resulting from new research efforts both in environmental sciences and in the areas of economic/ecologic interactions, eco-policy, eco-sociological problems, and eco-ethics.

Ready or not, coastal communities are facing a crisis with the anticipated population influx and the attendant problems it will bring. Decisions must be made to insure that optimum use of coastal resources be considered in tandem with minimal environmental degradation. In the decision-making process, it is important to learn what has been done in other areas, both mistakes and successes. To the extent possible, it works well to use existing environmental and natural resource management capabilities, and, as noted above, everyone who is interested should be included. The use of a regional planning framework and development of different scenarios for impact assessment should also be part of the process. It is not that the newest ideas and data will not, at times, lead us down the wrong path. Mistakes will be made. But by keeping an eye on the future, and remembering the past, while keeping sectors of society involved, the process of "smart growth" along the coast is a solution entirely within our grasp.

Notes

Chapter 1

1. Boorstin, D. *The discoverers.* New York: Random House, 1983.
2. Deacon, M. *Seas, maps, and men: an atlas-history of man's exploration of the oceans.* New York: Doubleday, 1962.
3. Ewing, G. From antiquity onward. *Oceanus* 24 (1981): 6–8.
4. Hull, S. *The bountiful sea.* London: Sidgwick and Jackson, 1964.
5. *Coastal Zone Management Act, 1994.* Oct. 27, 1972, P. L. 92-583, 86 Stat. 1280, 16. U.S. Code sections 1451 to 1464.
6. Ketchum, B. *The water's edge: critical problems of the coastal zone.* Cambridge, Mass.: Massachusetts Institute of Technology Press, 1972.
7. *South Carolina Tidelands Act of 1977.* S.C. Code Ann. 1976, sections 48-39-10, et seq.
8. United Nations. *Development of international economic and social affairs, 1990.* New York: United Nations, 1991.
9. Franz, D. An historical perspective on molluscs in lower New York harbor with emphasis on oysters. In Mayer, G., ed., *Ecological stress and the New York bight: science and management.* Columbia, S.C.: Belle W. Baruch Institute, University of South Carolina, 1982: 181–98.
10. Klee, G. *The coastal environment: toward integrated coastal and marine sanctuary management.* Engelwood Cliffs, N.J.: Prentice Hall, 1999.
11. *Conserving the nation's coasts and estuaries: a strategic plan for the national estuarine research system, a state and federal partnership.* Washington, D.C.: National Oceanic and Atmospheric Administration, 1995.

Chapter 2

1. Tansley, A. The use and abuse of vegetational concepts and terms. *Ecology* 16 (1935): 284–307.
2. Pandian, T., and Vernberg, F. *Animal Energetics.* Vols. 1 and 2. New York: Academic Press, 1983.
3. Limbaugh, C., Pederson H., and Chace, J. Shrimps that clean fishes. *Bulletin of Marine Science of the Gulf and Caribbean* 11 (1961): 237–57.
4. Vernberg, F., and Vernberg, W. *The animal and the environment.* New York: Holt, Rinehart and Winston, 1970.
5. Vernberg, W., and Vernberg, F. *Environmental physiology of marine animals.* New York, Heidelberg, and Berlin: Springer-Verlag, 1972.
6. Wells, H. The fauna of oyster beds, with special reference to the salinity factor. *Ecological Monographs* 31 (1961): 239–66.
7. Macmillan, D. *Tides.* New York: American Elsevier Publishing Company, 1966.
8. Summer, J., McKellar, J., Dame, R., and Kitchens, W. A simulation model of estuarine subsystem coupling and carbon exchange with the sea. Part

2, North Inlet model structure, output and validation. *Ecological Modeling* 11 (1980): 101–40.

9. Morris, J. Effects of nitrogen loading on wetland ecosystems with particular reference to atmospheric deposition. *Annual Review of Ecological Systems* 22 (1981): 257–79.

10. Vernberg, F. Salt-marsh processes: a review. *Environmental Toxicology and Chemistry* 12 (1993): 2167–95.

11. Likens, G. *Some perspectives of the major biogeochemical cycles.* New York: Wiley, 1981.

12. Krebs, C. *Ecology: the experimental analysis of distribution and abundance.* 3rd ed. New York: Macmillan, 1985.

13. Whittaker, R. *Communities and ecosystems.* New York: Macmillan, 1975.

14. Vernberg, F. Long-term ecological research on the North Inlet forest-wetlands-marine landscape, Georgetown, S.C. In *Barrier island/salt marsh estuaries, southeast Atlantic coast: issues, resources, status and management.* Washington, D.C.: U.S. Department of Commerce, 1989.

15. Dardeau, M., Modlin, R., Schroeder, W., and Stout, J. Estuaries. In Hackney, C., Adams, S., and Martin, W., eds. *Biodiversity of the southeastern United States.* New York: John Wiley & Sons, 1992: 615–744.

16. Turner, R. Geographic variations in salt marsh macrophyte production: a review. *Contributions to Marine Science* 20 (1976): 47–68.

17. Dame, R., and Kenny, P. Variability of *Spartina* alterniflora primary production in the euhaline North Inlet estuary. *Marine Ecology Progress Series* 32 (1986): 70–80.

18. Pomeroy, L., Darley, W., Dunn, E., Gallagher, J., Haines, E., and Whitney, D. Primary production. In Pomeroy, L. and Weigert, R., eds., *The Ecology of a Salt Marsh.* New York: Springer-Verlag, 1981: 39–67.

19. Dame, R., Zingmark, R., and Haskin, E. Oyster reefs as processors of estuarine materials. *Journal of Experimental Marine Biology and Ecology* 83 (1984): 239–47.

20. Davis, L. Class Insecta. In Zingmark, R., ed., *An annotated checklist of the biota of the coastal zone of South Carolina.* Columbia, S.C.: Belle W. Baruch Institute, University of South Carolina, 1978: 186–220.

21. Gunter, G., Ballard, B., and Venkataramiah, A. A review of salinity problems of organisms in United States coastal areas subject to the effects of engineering works. *Gulf Research Report* 1974: 380–475.

22. Barnes, R. *Estuarine Biology.* London: Edward Arnold, 1984.

23. Vernberg, W., DeCoursey, P., and O'Hara, J. Multiple environmental factor effects on physiology and behavior of the fiddler crab, *Uca pugilator.* In Vernberg, F., and Vernberg, W., eds., *Pollution and physiology of marine organisms,* 381–425. New York: Academic Press, 1974.

24. Pinet, P. *Oceanography: an introduction to the Planet Oceanus.* St. Paul: West Publishing Company, 1992.

25. Weigert, R., and Wetzel, R. Simulation experiments with a fourteen-compartment model of a *Spartina* salt marsh. In Dame, R., ed., *Marsh estuarine*

system simulation, 7–39. No. 8 of Belle W. Baruch Library in Marine Science. Columbia: University of South Carolina Press, 1979.

26. Meadows, P., and Campbell, J. *An introduction to marine science.* New York: Chapman & Hall, 1978.

27. Weiss, R. The solubility of nitrogen, oxygen, and argon in water and seawater. *Deep-Sea Research* 17 (1970): 721–35.

28. Kennish, M. *Pollution impacts on marine biotic communities.* Boca Raton, Fla.: CRC Press, 1997.

29. Merriam, D. Studies on the striped bass (*Roccus saxatilis*) of the Atlantic coast. *Fisheries Bulletin*, Unites States Fish and Wildlife Service, 50 (1974): 1–77.

Chapter 3

1. Hedgpeth, J. Classification of marine environments. In Hedgpeth, J., ed., *Marine ecology and paleoecology.* New York: *Geological Society of America Memoirs* 67 (1957): 17–28.

2. Pritchard, D. What is an estuary?: Physical viewpoint. In Lauff, G., ed., *Estuaries*, 3–5. Washington, D.C.: American Association for the Advancement of Science, 1967.

3. Gosselink, J., and Baumann, R. Wetland inventories: wetland loss along the United States coast. *Geomorphology Sppl.* (new series) 34 (1980): 173–87.

4. Chapman, V. Salt marshes and salt deserts of the world. In Reimold, R., and Queen, W., eds., *Ecology of halophytes*, 3–19. New York: Academic Press, 1974.

5. Redfield, A. Development of a New England salt marsh. *Ecological Monographs* 42 (1972): 210–37.

6. Fleesa, K., Constantine, K., and Kushman, M. Sedimentation rates in a coastal marsh determined from historical records. *Chesapeake Science* 18 (1977): 172–76.

7. Kjerfve, B., Greer, J., and Crout, L. Low frequency response of estuarine sea level to non-local forcing. In Wiley, M., ed., *Estuarine interactions.* New York: Academic Press, 1978.

8. Sharma, P., Gardner, L., Moore, W., and Bollinger, M. Sedimentation and bioturbation in a salt marsh as revealed by 210_{Pb}, 137_{Cs} and 7_{Bi} studies. *Limnology and Oceanography* 32 (1987): 313–26.

9. Morris, J., Kjerfve, B., and Dean, J. Dependence of estuarine productivity on anomalies in mean sea level. *Limnology and Oceanography* 35 (1990): 926–30.

10. Morris, J. Effects of nitrogen loading on wetland ecosystems with particular reference to atmospheric deposition. *Annual Review of Ecological Systems* 22 (1991): 257–79.

11. Vernberg, F., Vernberg, W., Blood, E., Fortner, A., Fulton, M., McKellar, H., Michener, W., Scott, G., Siewicki, T., and El-Figi, K. Impact of urbanization on high salinity estuaries in the southeastern United States. *Netherlands Journal of Sea Research* 30 (1992): 239–48.

12. Zingmark, R. *An annotated checklist of the biota of the coastal zone of South Carolina.* Columbia: University of South Carolina Press, 1978.

13. Gosselink, J. *The ecology of delta marshes of coastal Louisiana: a community*

profile. U.S. Fish and Wildlife Service Office, 1984; Technical Report FWS/OBS/84–09: 1–134.

14. Kneib, R. Patterns of invertebrate distribution and abundance in the intertidal salt marsh: cases and questions. *Estuaries* 7 (1984): 392–412.

15. Fox, R., and Ruppert, E. *Shallow-water marine benthic macroinvertebrates of South Carolina: species identification, community composition and symbiotic associations.* No. 14 of Belle W. Baruch Library in Marine Science. Columbia: University of South Carolina Press, 1985.

16. Ogburn, M., Allen, D., and Michener, W. *Fishes, shrimps, and crabs of the North Inlet Estuary, South Carolina: a four-year seine and trawl survey.* Columbia: University of South Carolina, 1988; Technical Report 88–1.

17. Ruppert, E., and Fox, R. *Seashore animals of the southeast.* Columbia: University of South Carolina Press, 1988.

18. Eleuterius, L. *Tidal marsh plants.* Gretna, La.: Pelican Publishing Company, 1990.

19. Remane, A., and Schlieper, C. *Biology of brackish water.* New York: John Wiley & Sons, 1971.

20. Vernberg, W., and Vernberg, F. *Environmental physiology of marine animals.* New York, Heidelberg, and Berlin: Springer-Verlag, 1972.

21. McLusky, D. *Tertiary level biology: the estuarine ecosystem.* New York: Halsted Press, 1981.

22. Hackney, C., Adams, S., and Martin, W. *Biodiversity of the southeastern United States aquatic communities.* New York: John Wiley & Sons, 1992.

23. Newell, R. *Biology of intertidal animals.* New York: American Elsevier Publishing Company, 1970.

24. Vernberg, F. *Physiological ecology of estuarine organisms.* No. 3 of Belle W. Baruch Library in Marine Science. Columbia: University of South Carolina Press, 1975.

25. Gilles, R. *Mechanisms of osmoregulation in animals.* New York: John Wiley & Sons, 1979.

26. Vernberg, F., and Vernberg, W. *Functional adaptations of marine organisms.* New York: Academic Press, 1981.

27. Pandian, T., and Vernberg, F. *Animal Energetics.* Vols. 1–2. New York: Academic Press, 1987.

28. Vernberg, F. Salt-marsh processes: a review. *Environmental Toxicology and Chemistry* 12 (1993): 2167–95.

29. Dame, R., Chrzanowski, T., Bildstein, K., Kjerfve, B., McKellar, H., Nelson, D., Spurrier, J., Stancyk, S., Stevenson, H., Vernberg, F., and Zingmark, R. The outwelling hypothesis and North Inlet, South Carolina. *Marine Ecology Progress Series* 33 (1986): 217–29.

30. Whiting, G., and Childers, D. Subtidal advective water flux as a potentially important nutrient input to southeastern USA salt marsh estuaries. *Estuarine, Coastal and Shelf Science* 28 (1989): 417–31.

31. Kjerfve, B., ed. *Caribbean coral reef, seagrass, and magrove sites.* Coastal Region and Small Island Papers. Paris: UNESCO, 1998.

32. Saenger, P., Hegerl, E., and Davie, J. Global status of mangrove ecosystems. *Environmentalist* 3 (1983): 1–88.

33. Hutchings, P., and Saenger, P. *Ecology of mangroves.* St. Lucia, Australia: University of Queensland Press, 1987.

34. Kjerfve, B., Lacerda, L., and Diop, E. *Mangrove ecosystems studies in Latin America and Africa.* Paris: UNESCO, 1997.

35. Yonge, C. The biology of coral reefs. *Advances in Marine Biology* 1 (1963): 209–60.

36. Lewis, J. Processes of organic production on coral reefs. *Biological Review* 52 (1977): 3305–47.

37. Sale, P. The ecology of fishes on coral reefs. *Oceanography and Marine Biology Annual Review* 18 (1980): 364–421.

38. Jones, O., and Endean, R. *The biology and geology of coral reefs.* New York: Academic Press, 1973.

39. Foster, M., and Schiel, D. *The ecology of giant kelp forests in California: a community profile.* U.S. Fish Wildlife Service Biological Report 85 (1985): 1–152.

40. North, W. The biology of giant kelp beds (*Macrocystis*) in California. *Nova Nedwigia* (1971): 600.

41. Tegner, M., Dayton, P., Edwards, P., et al. Effects of a large sewage spill on a kelp forest community: catastrophe or disturbance? *Marine Environmental Research* 40 (1995): 181–224.

Chapter 4

1. Katz, P. *The new urbanism: toward an architecture of community.* New York: McGraw-Hill, 1994.

2. Boorstin, D. *The discoverers.* New York: Random House, 1983.

3. Tiner, R. *Wetlands of the United States: current status and trends.* United States Fish and Wildlife Service, National Wetlands Inventory, Washington, D.C., 1984.

4. Shabman, L. Land settlement, public policy, and the environmental future of the southeast coast. In Vernberg, F., Vernberg, W., and Siewicki, T., eds., *Sustainable development in the southeastern coastal zone.* No. 20 of Belle W. Baruch Library in Marine Science. Columbia: University of South Carolina Press. 1996.

5. Taeuber, C. Population trends of the 1960s. *Science* 176 (1970): 773–77.

6. Culliton, T., Warren, M., Goodspeed, R., Remer, D., Blackwell, C., and J McDonough, I. *Fifty years of population change along the nation's coasts, 1960–2010.* Washington, D.C.: National Oceanic and Atmospheric Administration, Department of Commerce, 1990.

7. *Development of international economic and social affairs world urbanization prospects, 1990.* New York: United Nations, 1991.

8. Bailey, W. Population trends in the coastal area, concentrating on South Carolina. In Vernberg, F., Vernberg, W., and Siewicki, T., eds., *Sustainable development in the southeastern coastal zone.* No. 20 of Belle W. Baruch Library in Marine Science. Columbia: University of South Carolina Press, 1996.

9. Becker, R. *Some thoughts on coastal growth trends.* Southeast Coastal Ocean Research Conference, Savannah, Ga., 1998.

10. Downs, A. *The need for a new vision for the development of large U.S. metropolitan areas.* New York: Solomon Brothers, 1989.

11. Davis, J., Burch, J., Essex, D., Haeweler, M., Kinsey, M., Penney, J., and Schwartzkopf, W. Local policies to control coastal growth in the Waccamaw region. In Vernberg, F., Vernberg, W., and Siewicki, T., eds., *Sustainable development in the southeastern coastal zone.* No. 20 of Belle W. Baruch Library in Marine Science. Columbia: University of South Carolina Press, 1996.

12. Lubchenco, J., Olson, A., Brubaker, L., et al. The sustainable biosphere initiative: an ecological research agenda. *Ecology* 72 (1991): 371–412.

13. Fulton, M., Chandler, G., and Scott, G. Urbanization effects on the fauna of a southeastern USA bar-built estuary. In Vernberg, F., Vernberg, W., and Siewicki, T., eds., *Sustainable development in the southeastern coastal zone,* 477–504. No. 20 of Belle W. Baruch Library in Marine Science. Columbia: University of South Carolina Press, 1996.

14. Wahl, M., McKellar, H., Jr., and Williams, T. The effects of coastal development on watershed hydrography and the transport of organic carbon. In Vernberg, F., Vernberg, W., and Siewicki, T., eds., *Sustainable development in the southeastern coastal zone,* 378–412. No. 20 of Belle W. Baruch Library in Marine Science. Columbia: University of South Carolina Press, 1996.

15. Odum, E. Conceptual models relevant to sustaining coastal zone resources. In Vernberg, F., Vernberg, W., and Siewicki, T., eds., *Sustainable development in the southeastern coastal zone.* No. 20 of Belle W. Baruch Library in Marine Science. Columbia: University of South Carolina Press, 1996.

16. *Land development provisions to protect Georgia water quality.* Athens: Georgia Department of Natural Resources, 1997.

17. World Commission on Environment and Development. *Our common future.* Oxford: Oxford University Press, 1987.

18. Sinderman, C. Sustainable development in the southeastern coastal zone: a summary, 1997. In Vernberg, F., Vernberg, W., and Siewicki, T., eds., *Sustainable development in the southeastern coastal zone.* No. 20 of Belle W. Baruch Library in Marine Science. Columbia: University of South Carolina Press, 1996.

19. Arendt, R. *Rural by design: maintaining small town character.* Chicago: American Lanning Association. 1992.

20. Stuller, J. Golf looks beyond the "Augusta National Syndrome." *Smithsonian* 28 (1997): 5667.

21. Hurdzan, M. *Golf course architecture.* Chelsea, Mich.: Sleeping Bear Press, 1996.

Chapter 5

1. Winslow, E. *The evolution and significance of the modern public health campaign.* London: Oxford University Press, 1984.

2. Valette-Silver, N. The use of sediment cores to reconstruct historical

trends in contamination of estuarine and coastal sediments. *Estuaries* 16 (1993): 577–88.

3. O'Conner, T. Trends in chemical concentrations in mussels and oysters collected along the U.S. coast from 1993 to 1996. *Marine Environmental Research* 41 (1996): 183–200.

4. Weinstein, J. Anthropogenic impacts on salt marshes: a review. In Vernberg, F., Vernberg, W., and Siewicki, T., eds., *Sustainable development in the southeastern coastal zone*, 135–70. No. 20 of Belle W. Baruch Library in Marine Science. Columbia: University of South Carolina Press, 1996.

5. Waldichuk, M. Some biological concerns in heavy metals pollution. In Vernberg, F., and Vernberg, W., eds., *Pollution and physiology of marine organisms*. New York: Academic Press, 1974.

6. Kennish, M. *Ecology of estuaries: anthropogenic effects*. Boca Raton, Fla.: CRC Press, 1992.

7. Fortner, A., Sanders, M., and Lemire, S. Polynuclear aromatic hydrocarbon and trace metal burdens in sediment and the oyster, *Crassostrea virginica* Gmelin, from two high-salinity estuaries in South Carolina. In Vernberg, F., Vernberg, W., and Siewicki, T., eds., *Sustainable development in the southeastern coastal zone*, 445–76. No. 20 of Belle W. Baruch Library in Marine Science. Columbia: University of South Carolina Press, 1996.

8. Woodwell, G., Wurster, C., and Isaacson, P. DDT residues in an east coast estuary: a case of biological concentration of a persistent insecticide. *Science* 156 (1967): 821–24.

9. Elson, P. Effects on wild young salmon of spraying DDT over New Brunswick forests. *Journal of Fisheries Resesarch Board* 24 (1967): 731–67.

10. Long, M., Stephenson, M., Puckelt, M., Fairey, R., Hunt, J., Anderson, B., Holstad, D., Newman, J., and Birosik, S. Sediment chemistry and toxicity in the vicinity of the Los Angeles and Long Beach Harbors. Sacramento, Calif.: California State Water Resources Control Board, 1994.

11. Arnold, S., Klotz, D., Collins, B., Vonier, P., Jr. LG, and McLachlan, J. Synergistic activation of estrogen receptor with combinations of environmental chemicals. *Science* 272 (1996): 1489–92.

12. Colborn, T., Vom Saal, F., and Soto, A. Development effects of endocrine disrupting chemicals in wildlife and humans. *Environmental Health Perspectives* 101 (1993): 378–84.

13. Clark, R. *Marine pollution*. Oxford: Clarendon Press, 1989.

14. Carlberg, S. Oil pollution of the marine environment-with an emphasis on estuarine studies. In Olausson, E., and Cato, I., eds., *Chemistry and biogeochemistry of estuaries*. Chichester, U.K.: John Wiley and Sons, 1980: 367.

15. Percy, J., and Wells, P. Effects of petroleum in polar marine environments. *Marine Technology Society Journal* 18 (1984): 51–61.

16. Coull, B., and Chandler, G. Pollution and meiofauna: Field, laboratory and mesocosm studies. Oceanogr. *Marine Biology Annual Review* 30 (1992): 191–271.

17. Corner, E., Southward, A., and Southward, E. Toxicity of oil spill removers (detergents) to marine life: an assessment using the intertidal barnacle *Eliminius moclestus*. *Journal of Marine Biological Association* 48 (1968): 29–47.

18. Chandler, T., and Green, A. A 14–day harpacticoid copepod reproduction bioassay for laboratory and field-contaminated muddy sediments. In Ostrander, G., ed., *New techniques in aquatic toxicology*. Boca Raton, Fla.: Lewis Publishers, 1996: 23–39.

19. Bidleman, T. Personal communication, 1992.

20. Siewicki, T. Environmental modeling and exposure assessment of sedimentassociated fluoranthene in a small, urbanized, non-riverine estuary. *Journal of Experimental Marine Biology and Ecology* 213 (1996): 71–94.

21. Porter, D., Michener, W., Siewicki, T., Edwards, D., and Corbett, C. Geographic information processing assessment of the impacts of urbanization on localized coastal estuaries: a multidisciplinary approach. In Vernberg, F., Vernberg, W., and Siewicki, T., eds., *Sustainable development in the southeastern coastal zone*, 355–87. No. 20 of Belle W. Baruch Library in Marine Science. Columbia: University of South Carolina Press, 1996.

22. Long, E., and Markel, R. *An evaluation of the extent and magnitude of biological effects associated with chemical contaminants in San Francisco Bay, California. Seattle, Washington*. NOSORCA 64. National Oceanic and Atmospheric Administration Technical Memorandum, National Oceanic and Atmospheric Administration, 1992.

23. Chandler, G., Shipp, M., and Donelan, T. Bioaccumulation growth and larval settlement effects of sediment-associated polynuclear aromatic hydrocarbons on the estuarine polychaete, *Streblospio benedicti* (Webster). *Journal of Experimental Marine Biology and Ecology* (1996).

24. Long, E. The use of biological measures in assessments of toxicants in the coastal zone. In Vernberg, F., Vernberg, W., and Siewicki, T., eds., *Sustainable development in the southeastern coastal zone*. No. 20 of Belle W. Baruch Library in Marine Science. Columbia: University of South Carolina Press, 1996.

25. Fulton, M., Chandler, G., and Scott, G. Urbanization effects on the fauna of a southeastern USA Bar-built estuary. In Vernberg, F., Vernberg, W., and Siewicki, T., eds., *Sustainable development in the southeastern coastal zone*, 407–504. No. 20 of Belle W. Baruch Library in Marine Science. Columbia: University of South Carolina Press, 1996.

Chapter 6

1. O'Conner, T. Trends in chemical concentrations in mussels and oysters collected along the U.S. coast from 1993 to 1996. *Marine Environmental Research* 41 (1996): 183–200.

2. Anderson, D. Red tides. *Scientific American* 271 (1994): 52–58.

3. Anderson, D. *The ecology and oceanography of harmful algal bloom: A national research agenda*. Woods Hole, Mass.: Woods Hole Oceanographic Institution, 1995.

4. Tester, P., et al. *Gymnodinium breve* and global warning: What are the possibilities? In Smayda, T., and Shimizu, Y., eds., *Toxic phytoplankton blooms in the sea.* Amsterdam: Elsevier Science Publ. B. V., 1993: 67–72.

5. Franks, P., and Anderson, D. A longshore transport of a toxic phytoplankton bloom in a buoyancy current: *Alexandrium tamerense* in the Gulf of Maine. *Marine Biology* 112 (1992): 153–64.

6. Lam, C., and Ho, K. Red tides in Tolo Harbour, Hong Kong. In Lassus, P., Arzul, G., Denn, E., Gentien, P., and Marcaillou-Le Baut, C., eds., *Harmful Marine Algal Blooms.* Paris: Lavoisier Publishing, 1989: 65–70.

7. Shumway S. A review of the effects of algal blooms on shellfish and aquaculture. *Journal of World Aquaculture Society* 1990;21: 65–104.

8. Kahn, J., and Rochel, M. Measuring the economic effects of brown tides. *Journal of Shellfish Research* 7 (1988): 677–82.

9. Burkholder, J., Roya, J., Hobbs, C., and Glasgow, H., Jr. New 'phantom' dinoflagellate is the causative agent of major estuarine fish kills. *Nature* 358 (1992): 407–10.

10. Glasgow, H., Jr., and Burkholder, J. Insidious effects of a toxic estuarine dino-flagellate on fish survival and human health. *Journal of Toxicology and Environmental Health* 46 (1995): 501–22.

11. Burkholder, J., Glasgow, H., Jr., and Hobbs, C. Fish kills linked to a toxic ambush-predator dinoflagellate: distribution and environmental conditions. *Marine Ecology Progress Series* 124 (1995): 43–61.

12. Noga, E., Khoo, L., Stevens, J., Fan, Z., and Burkholder, J. Novel toxic dinoflagellate causes epidemic disease in estuarine fish. *Marine Pollution Bulletin* 32 (1995): 219–24.

13. Rublee, P. A., Kempton, J., Schaefer, E., Burkholder, J. M., Glasgow, H. B., and Oldach, D. PCR and FISH detection extends the rante of *Pfiesteria piscicida* in estuarine waters. *Virginia Journal of Science* 50 (1999) 325–35.

14. El-Figi, K. Epidemiological and microbiological evaluation of enteric bacterial waterborne diseases in coastal areas of South Carolina. Ph.D. diss., University of South Carolina, 1990.

15. Vernberg, W., Scott, G., Strozier, S., Bemiss, J., and Daugomah, J. The effects of urbanization on human and ecosystem health. In Vernberg, F., Vernberg, W., and Siewicki, T., eds., *Sustainable development in the southeastern coastal zone,* 221–39. No. 20 of Belle W. Baruch Library in Marine Science. Columbia: University of South Carolina Press, 1996.

16. Marcus, J. The impacts of selected land-use activities on the American oyster, *Crassostrea virginica* (Gmelin). Ph.D. diss., University of South Carolina, 1988.

17. Viraghaven, T., and Warnock, R. Groundwater quality adjacent to a septic tank system. *Journal of American Water Works Association* 68 (1976): 611–14.

18. DeWall, F., and Schaff, R. Ground-water pollution by septic tank drainfields. *Journal of Environmental Engineering Division Proceedings of American Society of Civil Engineers* 106 (1980): 631–46.

19. Scott, G., Sammons, T., Middaugh, D., and Hemmer, M. Impacts of

water chlorification and coliform bacteria on the American oyster, *Crassostrea virginica* (Gmelin). In Vernberg, W., Calabrese, A., Thurberg, F., and Vernberg, F., eds. *Physiological mechanisms of marine pollutant toxicity,* 505–529. New York: Academic Press, 1982.

20. Leonard, D., Slaughter, E., Genovase, P., Adamany, S., and Clement, C. *The 1990 national shellfish register of classified estuarine waters.* Rockville, Md.: National Oceanic and Atmospheric Administration, NOS/SAB, 1991.

21. Siewicki, T. Personal communication, 1999.

22. *American Water Works Association and Water Environment Federation Standard methods for the examination of water and wastewater.* 18th ed. Washington, D.C.: American Public Health Association. American Water Works Association, Water Environment Federation, 1992.

23. Watkins, W., and Burkhardt, W., III. New microbiological approaches for assessing and indexing contamination loading in estuaries and marine waters. In Vernberg, F., Vernberg, W., and Siewicki, T., eds., *Sustainable development in the southeastern coastal zone.* No. 20 of Belle W. Baruch Library in Marine Science. Columbia: University of South Carolina Press, 1996.

24. Rippey, S., and Watkins, W. Comparative rates of disinfection of microbial indicator organisms in chlorinated sewage effluents. *Water Science and Technology* 26 (1992): 2185–89.

25. Simmons, G.. Potential sources for non point introduction of *E. coli.* to tidal inlets. *Veterans Affairs Office of Coastal Resources Management* 23 (1997).

26. Scott, G. Personal communication, 1999.

27. Parveen, S., Murphree, R., Edmiston, L., Kaspar, C., Portier, K., and Tamplin, M. Association of multiple antibiotic resistance profiles with point and non point sources of *E coli* in Appalachicola Bay. *Applied and Environmental Microbiology* 63 (1997): 2607–12.

Chapter 7

1. VanDolah, R., Calder, D., and Knott, D. Effects of dredging and open-water disposal on benthic macroinvertebrates in a South Carolina estuary. *Estuaries* 7 (1984): 28–37.

2. Rhoads, D., and Boyer, L. The effects of marine benthos on physical properties of sediments: A successional perspective. In McCall, P., and Tevesz, M., eds., *Animal-Sediment Relations: The Biogenic Alteration of Sediments.* New York: Plenum Press, 1982: 3–52.

3. Holland, A., Porter, D., Dolah, R. V., Dunlap, R., Steel, G., and Upchurch, S. *Environmental assessment for alternative dredged material disposal sites in Charleston harbor.* Charleston, S.C.: South Carolina Wildlife and Marine Resources Department, Marine Resources Divisions, 1993.

4. Burger, J., and Shisler, J. Succession and productivity on perturbed and natural *Spartina* salt marsh areas in New Jersey. *Estuaries* 6 (1983): 50–56.

5. Weinstein, J. Anthropogenic impacts on salt marshes-a review. In Vernberg, F., Vernberg, W., and Siewicki, T., eds., *Sustainable development in the south-*

eastern coastal zone, 135–70. No. 20 of Belle W. Baruch Library in Marine Science. Columbia: University of South Carolina Press, 1996.

6. Wendt, P., Dolah, R. V., Bobo, M., and Manzi, J. *Effects of marina proximity on certain aspects of the biology of oysters and other benthic macrofauna in a South Carolina estuary.* Charleston, S.C.: South Carolina Wildlife and Marine Resources Department, Marine Resources Center, 1990.

7. Lennon, G., Neal, W., Bush, D., Pilkey, O., Stutz, M., and Bullock, J. *Living with the South Carolina Coast.* Durham, N.C.: Duke University Press, 1996.

8. *Coastal marinas assessment handbook.* Atlanta, Ga.: Environmental Protection Agency, 1985.

9. Ketchum, B. *The water's edge: critical problems of the coastal zone.* Cambridge, Mass.: Massachusetts Institute of Technology Press, 1972.

10. Pilkey, O. A "Thumbnail Method" for beach communities: estimation of long-term beach replenishment requirements. *Shore and Beach* (1988): 23–31.

11. Pilkey, O., and Clayton, T. Summary of beach replenishment experience on the U.S. East Coast Barrier Islands. *Journal of Coastal Research* 5 (1989): 147–59.

12. Broome, S., Seneca, E. W., and Woodhouse, J. Establishing brackish marshes on graded upland sites in North Carolina. *Wetlands* 2 (1982): 152–78.

13. Pilkey, O. H., and Wright, I. Seawalls versus beaches. *Journal of Coastal Research* 4 (1988): 41–64.

14. Chu, K., Tam, P., Fung, C., and Chen, Q. A biological survey of ballast water in container ships entering Hong Kong. *Hydrobiologia* 352 (1997): 201–6.

15. Kelly, J. Ballast water and sediments as mechanisms for unwanted species introductions into Washington State. *Journal of Shellfish Research* 12 (1993): 405–10.

16. Calvo-Ugarteburu, G., and McQuaid, C. Parasitism and introduced species: epidemiology of trematodes in the intertidal mussels *Perna perma* and *Mytilus. Journal of Experimental Marine Biology and Ecology* 220 (1998): 47–65.

17. Mills, C., and Sommer, F. Invertebrate introductions in marine habitats—2 species of hydromedusas (Cnidaria) native to the Black Sea, *Maeotias inexspectata* and *Blackfordia virginica*, invade San Francisco Bay. *Marine Biology* 122 (1995): 279–88.

18. Russell, D., and Balazs, G. Colonization by the alien marine alga *Hypnea musciformia* (Wulfen) J Ag (Rhodophyta, Gigartinales) in the Hawaiian Islands and its utilization by the green turtle, *Chelonia mydas* L. *Aquatic Botany* 47 (1994): 53–60.

19. Posey, M., Wigand, C., and Stevenson, J. Effects of an introduced aquatic plant, *Hydrilla verticillata*, on benthic communities in the upper Chesapeake Bay Estuarine. *Coastal Shelf Science* 37 (1993): 539–55.

20. Ramsay, K., Kaiser, J., and Hughes, R. Responses of benthic scavengers to fishing disturbance by towed gears in different habitats. *Journal of Experimental Marine Biology and Ecology* 224 (1998): 73–89.

21. Sykes, J. Estuarine use by oceanic finfish. In Newson, ed., *Proceedings of*

Marsh and Estuary Management Symposium. Baton Rouge: Louisiana State University, 1968.

22. Alexander, C., Brontman, M., and Field, D. *An inventory of coastal wetlands of the USA.* Washington, D.C.: NOAA, National Ocean Survey, U.S. Department of Commerce, 1986.

23. Broome, S., Seneca, E., and W. Woodhouse, J. Long-term growth and development of transplants of the salt marsh grass *Spartina alterniflora. Estuaries* 9 (1986): 214–21.

24. Broome, S., Seneca, E., and W. Woodhouse J. Tidal salt marsh restoration. *Aquatic Botany* 32 (1988): 1–22.

25. Matthews, G., and Minello, T. *Technology and success in restoration, creation, and enhancement of* Spartina alterniflora *marshes in the United States.* NOAA Coastal Ocean Program Decision Analysis Series No. 2. U.S. Department of Commerce, 1994.

26. Woodhouse, W., and Knutson, P. Atlantic coast marshes. In Lewis, R., ed. *Creation and restoration of coastal plan communities.* Boca Raton, Fla.: CRC Press, 1982.

27. Faber, P. The Muzzi marsh, Corte Madera, California, long-term observations of a restored marsh in San Francisco Bay, 424–38. In Bolton, H., ed., *Coastal wetlands.* New York: American Society of Civil Engineers, 1991.

28. LaSalle, M., Landin, M., and Sims, J. Evaluation of the flora and fauna of a *Spartina alterniflora* marsh established on dredged material in Winyah, South Carolina. *Wetlands* 11 (1991): 191–208.

29. Cammen, L. Macroinvertebrate colonization of *Spartina* marshes artificially established on dredge spoil. *Estuarine and Coastal Marine Science* 4 (1976): 357–72.

30. Cammen, L. Abundance and production of macroinvertebrates from natural and artificially established salt marshes in North Carolina. *American Midland Naturalist* 96 (1976): 487–93.

31. Craft, C., Broome, S., and Seneca, E. Nitrogen, phosphorus, and organic carbon pools in natural and transplanted marsh soils. *Estuaries* 11 (1988): 272–80.

32. Craft, C., Broome, S., and Seneca, E. Exchange of nitrogen, phosphorus, and organic carbon between transplanted marshes and estuarine waters. *Journal of Environmental Quality* 18 (1989): 206–11.

33. Brennan, J. Meteorological summary of Hurricane Hugo. *Journal of Coastal Research* 8 (1991): 1–12.

34. Sparks, P. Wind conditions in Hurricane Hugo and their effect on buildings in coastal South Carolina. *Journal of Coastal Research* 1991: 13–24.

35. Michener, W., Blood, E., Gardner, L., Kjerfve, B., Cabik, M., Coleman, C., Jefferson, W., Karinshak, D., and Spoons, F. *GIS assessment of large-scale ecological disturbances (Hurricane Hugo, 1989).* GIS/LIS 1991 Proceedings. Bethesda, Md.: American Congress Surveying and Mapping, 1991: 343–55.

36. Gardner, L., Michener, W., Blood, E., Williams, T., Lipscomb, D., and Jefferson, W. Ecological impact of Hurricane Hugo-salinization of a coastal forest. *Journal of Coastal Research* 8 (1991): 301–17.

37. Gardner, L., Michener, W., Williams, T., et al. Disturbance effects of Hurricane Hugo on a pristine coastal landscape: North Inlet, South Carolina, USA. *Netherlands Journal Sea Research* 30 (1992): 249–63.

38. Blood, E., Anderson, P., Smith, P., Nybro, C., and Ginsberg, K. Effects of Hurricane Hugo on coastal soil solution chemistry in South Carolina. *Biotropica* 23 (1991): 348–55.

39. Gardner, L., Michener, W., Kjerfve, B., and Karinshak, D. The geomorphic effects of Hurricane Hugo on an undeveloped coastal landscape at North Inlet, South Carolina. *Journal of Coastal Research* 8 (1991): 181–86.

40. Wells, H. The fauna of oyster beds, with special reference to the salinity factor. *Ecological Monographs* 31 (1961): 239–66.

41. Dunn, G. Some features of the hurricane problem. *Proceedings, National Shellfish Association*, 1956: 104–8.

42. Tibbetts, J. Hurricanes steer the course of history. *Coastal Heritage* 13 (1998): 3–9.

43. Hartmann, D. *Global physical climatology.* San Diego: Academic Press, 1994.

44. Houghton, J., ed. *Climate change 1995: the science of climate change.* Cambridge, Mass.: Cambridge University Press, 1996.

45. Titus, J., and Narayanan, V. *The probability of sea level rise.* Washington, D.C.: U.S. Environmental Protection Agency, 1995.

46. Emery, K. The continental shelves. *Scientific American* 221 (1969): 106–22.

47. Nicholls, R., and Leatherman, S. Potential impacts of accelerated sea level rise on developing countries. *Journal of Coastal Research* 14 (1995): 323.

48. Ervik, A., Hansen, P., Aure, J., Stingebrandt, A., Johannessen, P., and Jahnsen, T. Regulating the local environmental impact of intensive marine farming. Part 1. The concept of the MOM system (Modeling Ongrowing fish farms Monitoring). *Aquaculture* 158 (1977): 85–94.

49. Stevens, W. After the storm, an ecological bomb. *New York Times*, November 30, 1999.

50. Whitman, D. Hell in high water: Hurricane Floyd leaves behind an environmental nightmare. *U.S. News & World Report*, New York, October 4, 1999, p. 22.

51. Tester, P. Personal communication, 1999.

Chapter 8

1. National Environmental Policy Act of 1969. Jan. 1, 1970, P. L. 91-190, 83 Stat. 852, 42. U.S. Code sections 4321, 4331 to 4335, 4341 to 4347.

2. The Clean Air Act of 1972. Nov. 16, 1971, P. L. 92-157, 85 Stat. 464, 42. U.S. Code sections 1857c-8, 1857f-6c, 1857h-5.

3. Beatley, T., Brower, D., and Schwab, A. *An Introduction to Coastal Zone Management.* Washington, D.C.: Island Press, 1994.

4. Biliana, C., Knecht, R., Jang, D., and Fisk, G. *Integrated Coastal and Ocean Management: Concepts and Practices.* Washington, D.C.: Island Press, 1998.

5. Bird, E. *Beach Management*. New York: John Wiley and Sons, 1996.

6. Christie, D. *Coastal and Ocean Management in a Nutshell*. St. Paul: West Publishing Company, 1994.

7. Clark, J. *Coastal Zone Management Handbook*. New York: Lewis, 1996.

8. French, P. *Coastal and Estuarine Management*. New York: Routledge, 1997.

9. Hansom, J. *Coasts*. New York: Cambridge University Press, 1988.

10. Neal, W. *Living with the South Carolina Shore*. Durham: Duke University Press, 1984.

11. Pilkey, O., Neal, W., Bush, D., Shultz, M., Bulluck, J., and Lennon, G., eds., *Living with the South Carolina Coast*. Durham: Duke University Press, 1996.

12. Prince, H., and D'Intri, F. *Coastal Wetlands*. Chelsea, Mich.: Lewis Publishers, 1985.

Chapter 9

1. National Oceanic and Atmospheric Administration. *Conserving the nation's estuaries, a strategic plan for the National Estuarine Research Reserve System: a state and federal partnership*. Washington, D.C.: NOAA, 1995: 36.

2. National Research Council. *Priorities for coastal ecosystem science*. Washington, D.C.: National Academy Press, 1994: 79.

3. Michener, W., Lanter, D., and Houhoulis, P. Geographic information systems for sustainable development in the Southeastern United States: a review of applications and research needs. In Vernberg, F., Vernberg, W., and Siewicki, T., eds., *Sustainable development in the southeastern coastal zone*, 89–110. No. 20 of Belle W. Baruch Library in Marine Science. Columbia: University of South Carolina Press 1996.

4. Porter, D., Michener, W., Siewicki, T., Edwards, D., and Corbett, C. Geographic information processing assessment of the impacts of urbanization on localized coastal estuaries: a multidisciplinary approach. In Vernberg, F., Vernberg, W., and Siewicki, T., eds., *Sustainable development in the southeastern coastal zone*, 355–87. No. 20 of Belle W. Baruch Library in Marine Science. Columbia: University of South Carolina Press, 1996.

5. Long, E. The use of biological measures in assessments of toxicants in the coastal zone. In Vernberg, F., Vernberg, W., and Siewicki, T., eds., *Sustainable development in the southeastern coastal zone*, 187–220. No. 20 of Belle W. Baruch Library in Marine Science. Columbia: University of South Carolina Press, 1996.

6. Porter, D., and Salvesen, D., eds. *Collaborative planning for wetlands and wildlife*. Washington, D.C.: Island Press, 1995: 293.

7. Tibbetts, J. Investing in open space. *Coastal Heritage* 12, no. 4 (1998): 9–12. Charleston, S.C.: South Carolina Sea Grant Consortium.

8. Fodor, E. *Better not bigger*. Seattle: Island Press, 1999.

9. Coccossis, H., Burt, T., and van der Weide, J. Conceptual framework and planning guidelines for integrated coastal zone and river basin management. In Ozhan, E., ed. *Proceedings of the MEDCOAST 99–EMECS 99 Joint Conference:*

land ocean interactions—managing coastal ecosystems. Ankara, Turkey: Middle East Technical University, 1999: vol. 1, pp. 1–18.

10. Thomas, J. Changes in land cover in coastal areas and implications for fishery habitat. In Ozhan, E., ed., In Ozhan, E., ed. *Proceedings of the MED-COAST 99–EMECS 99 Joint Conference: land ocean interactions—managing coastal ecosystems.* Ankara, Turkey: Middle East Technical University, 1999: vol. 1, pp. 39–48.

11. Sinderman, C. Sustainable development in the southeastern coastal zone: a summary. In Vernberg, F., Vernberg, W., and Siewicki, T., eds., *Sustainable development in the southeastern coastal zone,* 509–17. No. 20 of Belle W. Baruch Library in Marine Science. Columbia: University of South Carolina Press, 1996.

Reference List

Alexander, C., Brontman, M., and Field, D. *An inventory of coastal wetlands of the USA.* Washington, D.C.: NOAA, National Ocean Survey, U.S. Department of Commerce, 1986.

American Water Works Association and Water Environment Federation Standard methods for the examination of water and wastewater. 18th ed. Washington, D.C.: American Public Health Association. American Water Works Association, Water Environment Federation, 1992.

Anderson, D. Red tides. *Scientific American* 271 (1994): 52–58.

Anderson, D. *The ecology and oceanography of harmful algal bloom: A national research agenda.* Woods Hole, Mass.: Woods Hole Oceanographic Institution, 1995.

Arendt, R. *Rural by design: maintaining small town character.* Chicago: American Lanning Association. 1992.

Arnold, S., Klotz, D., Collins, B., Vonier, P., Jr. LG, and McLachlan, J. Synergistic activation of estrogen receptor with combinations of environmental chemicals. *Science* 272 (1996): 1489–92.

Bailey, W. Population trends in the coastal area, concentrating on South Carolina. In Vernberg, F., Vernberg, W., and Siewicki, T., eds., *Sustainable development in the southeastern coastal zone.* No. 20 of Belle W. Baruch Library in Marine Science. Columbia: University of South Carolina Press, 1996.

Barnes, R. *Estuarine Biology.* London: Edward Arnold, 1984.

Beatley, T., Brower, D., and Schwab, A. *An Introduction to Coastal Zone Management.* Washington, D.C.: Island Press, 1994.

Becker, R. *Some thoughts on coastal growth trends.* Southeast Coastal Ocean Research Conference, Savannah, Ga., 1998.

Bidleman, T. Personal communication, 1992.

Biliana, C., Knecht, R., Jang, D., and Fisk, G. *Integrated Coastal and Ocean Management: Concepts and Practices.* Washington, D.C.: Island Press, 1998.

Bird, E. *Beach Management.* New York: John Wiley and Sons, 1996.

Blood, E., Anderson, P., Smith, P., Nybro, C., and Ginsberg, K. Effects of Hurricane Hugo on coastal soil solution chemistry in South Carolina. *Biotropica* 23 (1991): 348–55.

Boorstin, D. *The discoverers.* New York: Random House, 1983.

Brennan, J. Meteorological summary of Hurricane Hugo. *Journal of Coastal Research* 8 (1991): 1–12.

Broome, S., Seneca, E., and W. Woodhouse J. Tidal salt marsh restoration. *Aquatic Botany* 32 (1988): 1–22.

Broome, S., Seneca, E., and W. Woodhouse, J. Long-term growth and development of transplants of the salt marsh grass *Spartina alterniflora. Estuaries* 9 (1986): 214–21.

Broome, S., Seneca, E. W., and Woodhouse, J. Establishing brackish marshes on graded upland sites in North Carolina. *Wetlands* 2 (1982): 152–78.

Burger, J., and Shisler, J. Succession and productivity on perturbed and natural *Spartina* salt marsh areas in New Jersey. *Estuaries* 6 (1983): 50–56.

Burkholder, J., Roya, J., Hobbs, C., and Glasgow, H., Jr. New 'phantom' dinoflagellate is the causative agent of major estuarine fish kills. *Nature* 358 (1992): 407–10.

Burkholder, J., Glasgow, H., Jr., and Hobbs, C. Fish kills linked to a toxic ambush-predator dinoflagellate: distribution and environmental conditions. *Marine Ecology Progress Series* 124 (1995): 43–61.

Calvo-Ugarteburu, G., and McQuaid, C. Parasitism and introduced species: epidemiology of trematodes in the intertidal mussels *Perna perma* and *Mytilus. Journal of Experimental Marine Biology and Ecology* 220 (1998): 47–65.

Cammen, L. Abundance and production of macroinvertebrates from natural and artificially established salt marshes in North Carolina. *American Midland Naturalist* 96 (1976): 487–93.

Cammen, L. Macroinvertebrate colonization of *Spartina* marshes artificially established on dredge spoil. *Estuarine and Coastal Marine Science* 4 (1976): 357–72.

Carlberg, S. Oil pollution of the marine environment-with an emphasis on estuarine studies. In Olausson, E., and Cato, I., eds., *Chemistry and biogeochemistry of estuaries.* Chichester, U.K.: John Wiley and Sons, 1980: 367.

Chandler, T., and Green, A. A 14–day harpacticoid copepod reproduction bioassay for laboratory and field-contaminated muddy sediments. In Ostrander, G., ed., *New Techniques in Aquatic Toxicology.* Boca Raton, Fla.: Lewis Publishers, 1996: 23–39.

Chandler, G., Shipp, M., and Donelan, T. Bioaccumulation growth and larval settlement effects of sediment-associated polynuclear aromatic hydrocarbons on the estuarine polychaete, *Streblospio benedicti* (Webster). *Journal of Experimental Marine Biology and Ecology* 1996.

Chapman, V. Salt marshes and salt deserts of the world. In Reimold, R., and Queen, W., eds., *Ecology of halophytes,* 3–19. New York: Academic Press, 1974.

Christie, D. *Coastal and Ocean Management in a Nutshell.* St. Paul: West Publishing Company, 1994.

Chu, K., Tam, P., Fung, C., and Chen, Q. A biological survey of ballast water in container ships entering Hong Kong. *Hydrobiologia* 352 (1997): 201–6.

Clark, R. *Marine pollution.* Oxford: Clarendon Press, 1989.

Clark, J. *Coastal Zone Management Handbook.* New York: Lewis, 1996.

Coastal Marinas Assessment Handbook. Atlanta, Ga.: Environmental Protection Agency, 1985.

Coastal Zone Management Act, 1994. Oct. 27, 1972, P. L. 92-583, 86 Stat. 1280, 16. U.S. Code sections 1451 to 1464.

Coccossis, H., Burt, T., and van der Weide, J. Conceptual framework and planning guidelines for integrated coastal zone and river basin management. In Ozhan, E., ed. *Proceedings of the MEDCOAST 99–EMECS 99 Joint Conference: land ocean interactions—managing coastal ecosystems.* Ankara, Turkey: Middle East Technical University, 1999: vol. 1, pp. 1–18.

Colborn, T., Vom Saal, F., and Soto, A. Development effects of endocrine disrupting chemicals in wildlife and humans. *Environ Health Perspect* 101 (1993): 378–84.

Conserving the nation's coasts and estuaries: A strategic plan for the national estuarine research system A state and federal partnership. Washington, D.C.: National Oceanic and Atomospheric Administration, 1995.

Corner, E., Southward, A., and Southward, E. Toxicity of oil spill removers (detergents) to marine life: an assessment using the intertidal barnacle *Eliminius moclestus. Journal of Marine Biological Association* 48 (1968): 29–47.

Coull, B., and Chandler, G. Pollution and meiofauna: Field, laboratory and mesocosm studies. *Oceanographic and Marine Biology Annual Review* 30 (1992): 191–271.

Craft, C., Broome, S., and Seneca, E. Nitrogen, phosphorus, and organic carbon pools in natural and transplanted marsh soils. *Estuaries* 11 (1988): 272–80.

Craft, C., Broome, S., and Seneca, E. Exchange of nitrogen, phosphorus, and organic carbon between transplanted marshes and estuarine waters. *Journal of Environmental Quality* 18 (1989): 206–11.

Culliton, T., Warren, M., Goodspeed, R., Remer, D., Blackwell, C., and J McDonough, I. *Fifty years of population change along the nation's coasts, 1960–2010.* Washington, D.C.: National Oceanic and Atmospheric Administration, Department of Commerce, 1990.

Dame, R., and Kenny, P. Variability of *Spartina* alterniflora primary production in the euhaline North Inlet estuary. *Marine Ecology Progress Series* 32 (1986): 70–80.

Dame, R., Chrzanowski, T., Bildstein, K., Kjerfve, B., McKellar, H., Nelson, D., Spurrier, J., Stancyk, S., Stevenson, H., Vernberg, F., and Zingmark, R. The outwelling hypothesis and North Inlet, South Carolina. *Marine Ecology Progress Series* 33 (1986): 217–29.

Dame, R., Zingmark, R., and Haskin, E. Oyster reefs as processors of estuarine materials. *Journal of Experimental Marine Biology and Ecology* 83 (1984): 239–47.

Dardeau, M., Modlin, R., Schroeder, W., and Stout, J. Estuaries. In Hackney, C., Adams, S., and Martin, W., eds. *Biodiversity of the southeastern United States.* New York: John Wiley & Sons, 1992: 615–744.

Davis, J., Burch, J., Essex, D., Haeweler, M., Kinsey, M., Penney, J., and Schwartzkopf, W. Local policies to control coastal growth in the Waccamaw region. In Vernberg, F., Vernberg, W., and Siewicki, T., eds., *Sustainable development in the southeastern coastal zone.* No. 20 of Belle W. Baruch Library in Marine Science. Columbia: University of South Carolina Press, 1996.

Davis, L. Class Insecta. In Zingmark, R., ed., *An annotated checklist of the biota of the coastal zone of South Carolina.* Columbia, S.C.: Belle W. Baruch Institute, University of South Carolina, 1978: 186–220.

Deacon, M. *Seas, maps, and men: an atlas-history of man's exploration of the oceans.* New York: Doubleday, 1962.

Development of international economic and social affairs world urbanization prospects, 1990. New York: United Nations, 1991.

DeWall, F., and Schaff, R. Ground-water pollution by septic tank drainfields. *Journal of Environmental Engineering Division Proceedings of American Society of Civil Engineers* 106 (1980): 631–46.

Downs, A. *The need for a new vision for the development of large U.S. metropolitan areas.* New York: Solomon Brothers, 1989.

Dunn, G. Some features of the hurricane problem. *Proceedings, National Shellfish Association,* 1956: 104–8.

El-Figi, K. Epidemiological and microbiological evaluation of enteric bacterial waterborne diseases in coastal areas of South Carolina. Ph.D. diss., University of South Carolina, 1990.

Eleuterius, L. *Tidal marsh plants.* Gretna, La.: Pelican Publishing Company, 1990.

Elson, P. Effects on wild young salmon of spraying DDT over New Brunswick forests. *Journal of Fisheries Research Board* 24 (1967): 731–67.

Emery, K. The continental shelves. *Scientific American* 221 (1969): 106–22.

Ervik, A., Hansen, P., Aure, J., Stingebrandt, A., Johannessen, P., and Jahnsen, T. Regulating the local environmental impact of intensive marine farming. Part 1.

The concept of the MOM system (Modeling Ongrowing fish farms Monitoring). *Aquaculture* 158 (1977): 85–94.

Ewing, G. From antiquity onward. *Oceanus* 24 (1981): 6–8.

Faber, P. The Muzzi marsh, Corte Madera, California, long-term observations of a restored marsh in San Francisco Bay, 424–438. In Bolton, H., ed., *Coastal Wetlands.* New York: American Society of Civil Engineers, 1991.

Fleesa, K., Constantine, K., and Kushman, M. Sedimentation rates in a coastal marsh determined from historical records. *Chesapeake Science* 18 (1977): 172–76.

Fodor, E. *Better not bigger.* Seattle: Island Press, 1999.

Fortner, A., Sanders, M., and Lemire, S. Polynuclear aromatic hydrocarbon and trace metal burdens in sediment and the oyster, *Crassostrea virginica* Gmelin, from two high-salinity estuaries in South Carolina. In Vernberg, F., Vernberg, W., and Siewicki, T., eds., *Sustainable development in the southeastern coastal zone,* 445–76. No. 20 of Belle W. Baruch Library in Marine Science. Columbia: University of South Carolina Press, 1996.

Foster, M., and Schiel, D. *The ecology of giant kelp forests in California: a community profile.* U.S. Fish Wildlife Service Biological Report 85 (1985): 1–152.

Fox, R., and Ruppert, E. *Shallow-water marine benthic macroinvertebrates of South Carolina: species identification, community composition and symbiotic associations.* No. 14 of Belle W. Baruch Library in Marine Science. Columbia: University of South Carolina Press, 1985.

Franks, P., and Anderson, D. A longshore transport of a toxic phytoplankton bloom in a buoyancy current: *Alexandrium tamerense* in the Gulf of Maine. *Marine Biology* 112 (1992): 153–64.

Franz, D. An historical perspective on molluscs in lower New York harbor with emphasis on oysters. In Mayer, G. ed, *Ecological stress and the New York bight: science and management,* 181 98. Columbia, S.C.: Belle W. Baruch Institute for Marine Biology and Coastal Research, University of South Carolina, 1982.

French, P. *Coastal and Estuarine Management.* New York: Routledge, 1997.

Fulton, M., Chandler, G., and Scott, G. Urbanization effects on the fauna of a southeastern USA bar-built estuary. In Vernberg, F., Vernberg, W., and Siewicki, T., eds., *Sustainable development in the southeastern coastal zone,* 477–504. No. 20 of Belle W. Baruch Library in Marine Science. Columbia: University of South Carolina Press, 1996.

Gardner, L., Michener, W., Kjerfve, B., and Karinshak, D. The geomorphic effects of Hurricane Hugo on an undeveloped coastal landscape at North Inlet, South Carolina. *Journal of Coastal Research* 8 (1991): 181–86.

Gardner, L., Michener, W., Blood, E., Williams, T., Lipscomb, D., and Jefferson, W. Ecological impact of Hurricane Hugo-salinization of a coastal forest. *Journal of Coastal Research* 8 (1991): 301–17.

Gardner, L., Michener, W., Williams, T., et al. Disturbance effects of Hurricane Hugo on a pristine coastal landscape: North Inlet, South Carolina, USA. *Netherlands Journal Sea Research* 30 (1992): 249–63.

Gilles, R. *Mechanisms of osmoregulation in animals.* New York: John Wiley & Sons, 1979.

Glasgow, H., Jr., and Burkholder, J. Insidious effects of a toxic estuarine dino-flagellate on fish survival and human health. *Journal of Toxicology and Environmental Health* 46 (1995): 501–22.

Gosselink, J., and Baumann, R. Wetland inventories: wetland loss along the United States coast. *Geomorphology Supplement* (new series) 34 (1980): 173–87.

Gosselink, J. *The ecology of delta marshes of coastal Louisiana: A community profile.* U.S. Fish and Wildlife Service Office, 1984; Technical Report FWS/OBS/84–09: 1–134.

Gunter, G., Ballard, B., and Venkataramiah, A. A review of salinity problems of organisms in United States coastal areas subject to the effects of engineering works. *Gulf Research Report* 1974: 380–475.

Hackney, C., Adams, S., and Martin, W. *Biodiversity of the southeastern United States aquatic communities.* New York: John Wiley & Sons, 1992.

Hansom, J. *Coasts.* New York: Cambridge University Press, 1988.

Hartmann, D. *Global physical climatology.* San Diego: Academic Press, 1994.

Hedgpeth, J. Classification of marine environments. In Hedgpeth, J., ed., *Marine ecology and paleoecology.* New York: *Geological Society of America Memoirs* 67 (1957): 17–28.

Holland, A., Porter, D., Dolah, R., Dunlap, R., Steel, G., and Upchurch, S. *Environmental assessment for alternative dredged material disposal sites in Charleston harbor.* Charleston, S.C.: South Carolina Wildlife and Marine Resources Department, Marine Resources Divisions, 1993.

Houghton, J., ed. *Climate change 1995: the science of climate change.* Cambridge, Mass.: Cambridge University Press, 1996.

Hull, S. *The bountiful sea.* London: Sidgwick and Jackson, 1964.

Hurdzan, M. *Golf course architecture.* Chelsea, Mich.: Sleeping Bear Press, 1996.

Hutchings, P., and Saenger, P. *Ecology of mangroves.* St. Lucia, Australia: University of Queensland Press, 1987.

Jones, O., and Endean, R. *The biology and geology of coral reefs.* New York: Academic Press, 1973.

Kahn, J., and Rochel, M. Measuring the economic effects of brown tides. *Journal of Shellfish Research* 7 (1988): 677–82.

Katz, P. *The new urbanism: toward an architecture of community.* New York: McGraw-Hill, 1994.

Kelly, J. Ballast water and sediments as mechanisms for unwanted species introductions into Washington State. *Journal of Shellfish Research* 12 (1993): 405–10.

Kennish, M. *Ecology of estuaries: anthropogenic effects.* Boca Raton, Fla.: CRC Press, 1992.

Kennish, M. *Pollution impacts on marine biotic communities.* Boca Raton, Fla.: CRC Press, 1997.

Ketchum, B. *The water's edge: critical problems of the coastal zone.* Cambridge, Mass.: Massachusetts Institute of Technology Press, 1972.

Kjerfve, B., Lacerda, L., and Diop, E. *Mangrove ecosystems studies in Latin America and Africa.* Paris: UNESCO, 1997.

Kjerfve, B., ed. *Caribbean coral reef, seagrass, and mangrove sites.* Coastal Region and Small Island Papers. Paris: UNESCO, 1998.

Kjerfve, B., Greer, J., and Crout, L. Low frequency response of estuarine sea level to non-local forcing. In Wiley, M., ed., *Estuarine interactions.* New York: Academic Press, 1978.

Klee, G. *The coastal environment: toward integrated coastal and marine sanctuary management.* Engelwood Cliffs, N.J.: Prentice Hall, 1999.

Kneib, R. Patterns of invertebrate distribution and abundance in the intertidal salt marsh: cases and questions. *Estuaries* 7 (1984): 392–412.

Krebs, C. *Ecology: the experimental analysis of distribution and abundance.* 3rd ed. New York: Macmillan, 1985.

Lam, C., and Ho, K. Red tides in Tolo Harbour, Hong Kong. In Lassus, P., Arzul, G., Denn, E., Gentien, P., and Marcaillou-Le Baut, C., eds., *Harmful Marine Algal Blooms.* Paris: Lavoisier Publishing, 1989: 65–70.

Land development provisions to protect Georgia water quality. Athens: Georgia Department of Natural Resources, 1997.

LaSalle, M., Landin, M., and Sims, J. Evaluation of the flora and fauna of a *Spartina alterniflora* marsh established on dredged material in Winyah, South Carolina. *Wetlands* 11 (1991): 191–208.

Lennon, G., Neal, W., Bush, D., Pilkey, O., Stutz, M., and Bullock, J. *Living with the South Carolina coast.* Durham, N.C.: Duke University Press, 1996.

Leonard, D., Slaughter, E., Genovase, P., Adamany, S., and Clement, C. *The 1990 national shellfish register of classified estuarine waters.* Rockville, Md.: National Oceanic and Atmospheric Administration, NOS/SAB, 1991.

Lewis, J. Processes of organic production on coral reefs. *Biological Review* 52 (1977): 3305–47.

Likens, G. *Some perspectives of the major biogeochemical cycles.* New York: Wiley, 1981.

Limbaugh, C., Pederson H., and Chace, J. Shrimps that clean fishes. *Bulletin of Marine Science of the Gulf and Caribbean* 11 (1961): 237–257.

Long, M., Stephenson, M., Puckelt, M., Fairey, R., Hunt, J., Anderson, B., Holstad, D., Newman, J., and Birosik, S. Sediment chemistry and toxicity in the vicinity of the Los Angeles and Long Beach Harbors. Sacramento, Calif.: California State Water Resources Control Board, 1994.

Long, E., and Markel, R. *An evaluation of the extent and magnitude of biological effects associated with chemical contaminants in San Francisco Bay, California. Seattle, Washington.* NOSORCA 64. National Oceanic and Atmospheric Administration Technical Memorandum National Oceanic and Atmospheric Administration, 1992.

Long, E. The use of biological measures in assessments of toxicants in the coastal zone. In Vernberg, F., Vernberg, W., and Siewicki, T., eds., *Sustainable development in the southeastern coastal zone,* 187–220. No. 20 of Belle W. Baruch Library in Marine Science. Columbia: University of South Carolina Press, 1996.

Lubchenco, J., Olson, A., Brubaker, L., et al. The sustainable biosphere initiative: an ecological research agenda. *Ecology* 72 (1991): 371–412.

Macmillan, D. *Tides.* New York: American Elsevier Publishing Company, 1966.

Marcus, J. The impacts of selected land-use activities on the American oyster, *Crassostrea virginica (Gmelin).* Ph.D. diss., University of South Carolina, 1988.

Matthews, G., and Minello, T. *Technology and success in restoration, creation, and enhancement of* Spartina alterniflora *marshes in the United States.* NOAA Coastal Ocean Program Decision Analysis Series No. 2. U.S. Department of Commerce, 1994.

McLusky, D. *Tertiary level biology, the estuarine ecosystem.* New York: Halsted Press, 1981.

Meadows, P., and Campbell, J. *An introduction to marine science.* New York: Chapman & Hall, 1978.

Merriam, D. Studies on the striped bass (*Roccus saxatilis*) of the Atlantic coast. *Fisheries Bulletin* 50 (1974): 1–77.

Michener, W., Lanter, D., and Houhoulis, P. Geographic information systems for sustainable development in the Southeastern United States: a review of applications and research needs. In Vernberg, F., Vernberg, W., and Siewicki, T., eds., *Sustainable development in the southeastern coastal zone*, 89–110. No. 20 of Belle W. Baruch Library in Marine Science. Columbia: University of South Carolina Press 1996.

Michener, W., Blood, E., Gardner, L., Kjerfve, B., Cabik, M., Coleman, C., Jefferson, W., Karinshak, D., and Spoons, F. *GIS assessment of large-scale ecological disturbances (Hurricane Hugo, 1989)*. GIS/LIS 1991 Proceedings. Bethesda, Md.: American Congress Surveying and Mapping, 1991: 343–55.

Mills, C., and Sommer, F. Invertebrate introductions in marine habitats—2 species of hydromedusas (Cnidaria) native to the Black Sea, *Maeotias inexspectata* and *Blackfordia virginica*, invade San Francisco Bay. *Marine Biology* 122 (1995): 279–88.

Morris, J. Effects of nitrogen loading on wetland ecosystems with particular reference to atmospheric deposition. *Annual Review of Ecological Systems* 22 (1991): 257–79.

Morris, J., Kjerfve, B., and Dean, J. Dependence of estuarine productivity on anomalies in mean sea level. *Limnology and Oceanography* 35 (1990): 926–30.

Morris, J. Effects of nitrogen loading on wetland ecosystems with particular reference to atmospheric deposition. *Annual Review of Ecological Systems* 22 (1981): 257–79.

National Research Council. *Priorities for coastal ecosystem science*. Washington, D.C.: National Academy Press, 1994: 79.

National Environmental Policy Act of 1969. Jan. 1, 1970, P. L. 91-190, 83 Stat. 852, 42. U.S. Code sections 4321, 4331 to 4335, 4341 to 4347.

National Oceanic and Atmospheric Administration. *Conserving the nation's estuaries: a strategic plan for the National Estuarine Research Reserve System, a state and federal partnership*. Washington, D.C.: NOAA, 1995: 36.

Neal, W. *Living with the South Carolina shore*. Durham: Duke University Press, 1984.

Newell, R. *Biology of intertidal animals*. New York: American Elsevier Publishing Company, 1970.

Nicholls, R., and Leatherman, S. Potential impacts of accelerated sea level rise on developing countries. *Journal of Coastal Research* 14 (1995): 323.

Noga, E., Khoo, L., Stevens, J., Fan, Z., and Burkholder, J. Novel toxic dinoflagellate causes epidemic disease in estuarine fish. *Marine Pollution Bulletin* 32 (1995): 219–24.

North, W. The biology of giant kelp beds (*Macrocystis*) in California. *Nova Nedwigia* (1971): 600.

O'Conner, T. Trends in chemical concentrations in mussels and oysters collected along the U.S. coast from 1993 to 1996. *Marine Environmental Research* 41 (1996): 183–200.

Odum, E. Conceptual models relevant to sustaining coastal zone resources. In Vernberg, F., Vernberg, W., and Siewicki, T., eds., *Sustainable development in the southeastern coastal zone*, 75 89. No. 20 of Belle W. Baruch Library in Marine Science. Columbia: University of South Carolina Press, 1996.

Ogburn, M., Allen, D., and Michener, W. *Fishes, shrimps, and crabs of the North Inlet Estuary, South Carolina: a four-year seine and trawl survey*. Technical Report 88–1. Columbia: University of South Carolina, 1988.

Pandian, T., and Vernberg, F. *Animal Energetics*. Vols. 1 and 2. New York: Academic Press, 1983.

Parveen, S., Murphree, R., Edmiston, L., Kaspar, C., Portier, K., and Tamplin, M.

Association of multiple antibiotic resistance profiles with point and non-point sources of *E coli* in Appalachicola Bay. *Applied and Environmental Microbiology* 63 (1997): 2607–12.

Percy, J., and Wells, P. Effects of petroleum in polar marine environments. *Marine Technology Society Journal* 18 (1984): 51–61.

Pilkey, O. A "thumbnail method" for beach communities: estimation of long-term beach replenishment requirements. *Shore and Beach* (1988): 23–31.

Pilkey, O., and Clayton, T. Summary of beach replenishment experience on the U.S. East Coast Barrier Islands. *Journal of Coastal Research* 5 (1989): 147–59.

Pilkey, O., and Wright, I. Seawalls versus beaches. *Journal of Coastal Research* 4 (1988): 41–64.

Pilkey, O., Neal, W., Bush, D., Shultz, M., Bulluck, J., and Lennon, G., eds., *Living with the South Carolina coast.* Durham: Duke University Press.

Pinet, P. *Oceanography: an introduction to the Planet Oceanus.* St. Paul: West Publishing Company, 1992.

Pomeroy, L., Darley, W., Dunn, E., Gallagher, J., Haines, E., and Whitney, D. Primary production, 39–67. In Pomeroy, L. and Weigert, R., eds., *The Ecology of a Salt Marsh.* New York: Springer-Verlag, 1981.

Porter, D., Michener, W., Siewicki, T., Edwards, D., and Corbett, C. Geographic information processing assessment of the impacts of urbanization on localized coastal estuaries: a multidisciplinary approach. In Vernberg, F., Vernberg, W., and Siewicki, T., eds., *Sustainable development in the southeastern coastal zone,* 355–87. No. 20 of Belle W. Baruch Library in Marine Science. Columbia: University of South Carolina Press, 1996.

Porter, D., and Salvesen, D., eds. *Collaborative planning for wetlands and wildlife.* Washington, D.C.: Island Press, 1995: 293.

Porter, D., Michener, W., Siewicki, T., Edwards, D., and Corbett, C. Geographic information processing assessment of the impacts of urbanization on localized coastal estuaries: a multidisciplinary approach. In Vernberg, F., Vernberg, W., and Siewicki, T., eds., *Sustainable development in the southeastern coastal zone,* 496–507. No. 20 of Belle W. Baruch Library in Marine Science. Columbia: University of South Carolina Press, 1996.

Posey, M., Wigand, C., and Stevenson, J. Effects of an introduced aquatic plant, *Hydrilla verticillata,* on benthic communities in the upper Chesapeake Bay Estuarine. *Coastal Shelf Science* 37 (1993): 539–55.

Prince, H., and D'Intri, F. *Coastal Wetlands.* Chelsea, Mich.: Lewis Publishers, 1985.

Pritchard, D. What is an estuary?: Physical viewpoint. In Lauff, G., ed., *Estuaries,* 3–5. Washington, D.C.: American Association for the Advancement of Science, 1967.

Ramsay, K., Kaiser, J., and Hughes, R. Responses of benthic scavengers to fishing disturbance by towed gears in different habitats,. *Journal of Experimental Marine Biology and Ecology* 224 (1998): 73–89.

Redfield, A. Development of a New England salt marsh. *Ecological Monographs* 42 (1972): 210–37.

Remane, A., and Schlieper, C. *Biology of brackish water.* New York: John Wiley & Sons, 1971.

Rhoads, D., and Boyer, L. The effects of marine benthos on physical properties of sediments: A successional perspective. In McCall, P., and Tevesz, M., eds., *Animal-Sediment Relations: The Biogenic Alteration of Sediments.* New York: Plenum Press, 1982: 3–52.

Rippey, S., and Watkins, W. Comparative rates of disinfection of microbial indicator organisms in chlorinated sewage effluents. *Water Science and Technology* 26 (1992): 2185–89.

Rublee, P. A., Kempton, J., Schaefer, E., Burkholder, J. M., Glasgow, H. B., and Oldach, D. PCR and FISH detection extends the rante of *Pfiesteria piscicida* in estuarine waters. *Virginia Journal of Science* 50 (1999) 325–35.

Ruppert, E., and Fox, R. *Seashore animals of the southeast.* Columbia: University of South Carolina Press, 1988.

Russell, D., and Balazs, G. Colonization by the alien marine alga *Hypnea musciformia* (Wulfen) J Ag (Rhodophyta, Gigartinales) in the Hawaiian Islands and its utilization by the green turtle, *Chelonia mydas* L. *Aquatic Botany* 47 (1994): 53–60.

Saenger, P., Hegerl, E., and Davie, J. Global status of mangrove ecosystems. *Environmentalist* 3 (1983): 1–88.

Sale, P. The ecology of fishes on coral reefs. *Oceanographic and Marine Biology Annual Review* 18 (1980): 364–421.

Scott, G. Personal communication, 1999.

Scott, G., Sammons, T., Middaugh, D., and Hemmer, M. Impacts of water chlorification and coliform bacteria on the American oyster, *Crassostrea virginica* (Gmelin). In Vernberg, W., Calabrese, A., Thurberg, F., and Vernberg, F., eds. *Physiological mechanisms of marine pollutant toxicity*, 505–29. New York: Academic Press, 1982.

Shabman, L. Land settlement, public policy, and the environmental future of the southeast coast. In Vernberg, F., Vernberg, W., and Siewicki, T., eds., *Sustainable development in the southeastern coastal zone*. No. 20 of Belle W. Baruch Library in Marine Science. Columbia: University of South Carolina Press, 1996.

Sharma, P., Gardner, L., Moore, W., and Bollinger, M. Sedimentation and bioturbation in a salt marsh as revealed by 210_{Pb}, 137_{Cs} and 7_{Bi} studies. *Limnology and Oceanography* 32 (1987): 313–26.

Shumway, S. A review of the effects of algal blooms on shellfish and aquaculture. *Journal of World Aquaculture Society* 21 (1990): 65–104.

Siewicki, T. Environmental modeling and exposure assessment of sedimentassociated fluoranthene in a small, urbanized, non-riverine estuary. *Journal of Experimental Marine Biology and Ecology* 213 (1996): 71–94.

Siewicki, T. Personal communication, 1999.

Simmons, G.. Potential sources for nonpoint introduction of *E. coli.* to tidal inlets. *Veterans Affairs Office of Coastal Resources Management* 23 (1997).

Sinderman, C. Sustainable development in the southeastern coastal zone: a summary. In Vernberg, F., Vernberg, W., and Siewicki, T., eds., *Sustainable development in the southeastern coastal zone*, 509–17. No. 20 of Belle W. Baruch Library in Marine Science. Columbia: University of South Carolina Press, 1996.

Sinderman, C. Sustainable development in the southeastern coastal zone: a summary, 1997. In Vernberg, F., Vernberg, W., and Siewicki, T., eds., *Sustainable development in the southeastern coastal zone*. No. 20 of Belle W. Baruch Library in Marine Science. Columbia: University of South Carolina Press, 1996.

South Carolina Tidelands Act of 1977. S. C. Code Ann. 1976, sections 48-39-10 et seq.

Sparks, P. Wind conditions in Hurricane Hugo and their effect on buildings in coastal South Carolina. *Journal of Coastal Research* 1991: 13–24.

Stevens, W. After the storm, an ecological bomb. *New York Times*, November 30, 1999.

Stuller, J. Golf looks beyond the "Augusta National Syndrome." *Smithsonian* 28 (1997): 5667.

Summer, J., McKellar, J., Dame, R., and Kitchens, W. A simulation model of estuarine subsystem coupling and carbon exchange with the sea. Part 2, North Inlet model structure, output and validation. *Ecological Modeling* 11 (1980): 101–40.

Sykes, J. Estuarine use by oceanic finfish. In Newson, ed., *Proceedings of Marsh and Estuary Management Symposium*. Baton Rouge: Louisiana State University, 1968.

Taeuber, C. Population trends of the 1960s. *Science* 176 (1970): 773–77.

Tansley, A. The use and abuse of vegetational concepts and terms. *Ecology* 16 (1935): 284–307.

Tegner, M., Dayton, P., Edwards, P., et al. Effects of a large sewage spill on a kelp forest community: catastrophe or disturbance? *Marine Environmental Research* 40 (1995): 181–224.

Tester, P., et al. *Gymnodinium breve* and global warning: What are the possibilities? In Smayda, T., and Shimizu, Y., eds., *Toxic phytoplankton blooms in the sea*. Amsterdam: Elsevier Science Publ. B. V., 1993: 67–72.

Tester, P. Personal communication, 1999.

The Clean Air Act of 1972. Washington, D.C. United States Congress.

Thomas, J. Changes in land cover in coastal areas and implications for fishery habitat. In Ozhan, E., ed., In Ozhan, E., ed. *Proceedings of the MEDCOAST 99–EMECS 99 Joint Conference: land ocean interactions—managing coastal ecosystems*. Ankara, Turkey: Middle East Technical University, 1999: vol. 1, pp. 39–48.

Tibbetts, J. Hurricanes steer the course of history. *Coastal Heritage* 13 (1998): 3–9.

Tibbetts, J. Investing in open space. *Coastal Heritage* 12, no. 4 (1998): 9–12. Charleston, S.C.: South Carolina Sea Grant Consortium.

Tiner, R. *Wetlands of the United States: current status and trends*. United States Fish and Wildlife Service, National Wetlands Inventory, Washington, D.C., 1984.

Titus, J., and Narayanan, V. *The probability of sea level rise*. Washington, D.C.: U.S. Environmental Protection Agency, 1995.

Turner, R. Geographic variations in salt marsh macrophyte production: a review. *Contributions to Marine Science* 20 (1976): 47–68.

United Nations. *Development of international economic and social affairs, 1990*. New York: United Nations, 1991.

Valette-Silver, N. The use of sediment cores to reconstruct historical trends in contamination of estuarine and coastal sediments. *Estuaries* 16 (1993): 577–88.

VanDolah, R., Calder, D., and Knott, D. Effects of dredging and open-water disposal on benthic macroinvertebrates in a South Carolina estuary. *Estuaries* 7 (1984): 28–37.

Vernberg, F. Salt-marsh processes: a review. *Environmental Toxicology and Chemistry* 12 (1993): 2167–95.

Vernberg, F. *Physiological ecology of estuarine organisms*. No. 3 of Belle W. Baruch Library in Marine Science. Columbia: University of South Carolina Press, 1975.

Vernberg, W., DeCoursey, P., and O'Hara, J. Multiple environmental factor effects on physiology and behavior of the fiddler crab, *Uca pugilator*. In Vernberg, F., and Vernberg, W., eds., *Pollution and physiology of marine organisms*, 381–425. New York, Academic Press, 1974.

Vernberg, F., Vernberg, W., Blood, E., Fortner, A., Fulton, M., McKellar, H., Michener, W., Scott, G., Siewicki, T., and El-Figi, K. Impact of urbanization on high salinity estuaries in the southeastern United States. *Netherlands Journal of Sea Research* 30 (1992): 239–48.

Vernberg, W., Scott, G., Strozier, S., Bemiss, J., and Daugomah, J. The effects of urbanization on human and ecosystem health. In Vernberg, F., Vernberg, W., and Siewicki, T., eds., *Sustainable development in the southeastern coastal zone*, 221–39. No. 20 of Belle W. Baruch Library in Marine Science. Columbia: University of South Carolina Press, 1996.

Vernberg, F., and Vernberg, W. *Functional adaptations of marine organisms*. New York: Academic Press, 1981.

Vernberg, F., and Vernberg, W. *The animal and the environment*. New York: Holt, Rinehart and Winston, 1970.

Vernberg, F. Salt-marsh processes: a review. *Environmental Toxicology and Chemistry* 12 (1993): 2167–95.

Vernberg, F. *Long-term ecological research on the North Inlet forest-wetlands-marine landscape, Georgetown, S.C.* In *Barrier island/salt marsh estuaries, southeast Atlantic coast: issues, resources, status and management*. Washington, D.C.: U.S. Department of Commerce, 1989.

Vernberg, W and Vernberg, F. *Environmental physiology of marine animals*. New York, Heidelberg and Berlin: Springer-Verlag, 1972.

Viraghaven, T., and Warnock, R. Groundwater quality adjacent to a septic tank system. *Journal of American Water Works Association* 68 (1976): 611–14.

Wahl, M., McKellar, H., Jr, and Williams, T. The effects of coastal development on watershed hydrography and the transport of organic carbon. In Vernberg, F., Vernberg, W., and Siewicki, T., eds., *Sustainable development in the southeastern coastal zone*, 378–412. No. 20 of Belle W. Baruch Library in Marine Science. Columbia: University of South Carolina Press, 1996.

Waldichuk, M. Some biological concerns in heavy metals pollution. In Vernberg, F., and Vernberg, W., eds., *Pollution and physiology of marine organisms*. New York: Academic Press, 1974.

Watkins, W., and Burkhardt, W., III. New microbiological approaches for assessing and indexing contamination loading in estuaries and marine waters. In Vernberg, F., Vernberg, W., and Siewicki, T., eds., *Sustainable development in the southeastern coastal zone*. No. 20 of Belle W. Baruch Library in Marine Science. Columbia: University of South Carolina Press, 1996.

Weigert, R., and Wetzel, R. Simulation experiments with a fourteen-compartment model of a *Spartina* salt marsh. In Dame, R., ed., *Marsh estuarine system simulation*, 7–39. No. 8 of Belle W. Baruch Library in Marine Science. Columbia: University of South Carolina Press, 1979.

Weinstein, J. Anthropogenic impacts on salt marshes-a review. In Vernberg, F., Vernberg, W., and Siewicki, T., eds., *Sustainable development in the southeastern coastal zone*, 135–70. No. 20 of Belle W. Baruch Library in Marine Science. Columbia: University of South Carolina Press, 1996.

Weiss, R. The solubility of nitrogen, oxygen, and argon in water and seawater. *Deep-Sea Research* 17 (1970): 721–35.

Wells, H. The fauna of oyster beds, with special reference to the salinity factor. *Ecological Monographs* 31 (1961): 239–66.

Wendt, P., Dolah, R. V., Bobo, M., and Manzi, J. *Effects of marina proximity on certain aspects of the biology of oysters and other benthic macrofauna in a South Carolina estuary*. Charleston, S.C.: South Carolina Wildlife and Marine Resources Department, Marine Resources Center, 1990.

Whiting, G., and Childers, D. Subtidal advective water flux as a potentially important nutrient input to southeastern USA salt marsh estuaries. *Estuarine, Coastal and Shelf Science* 28 (1989): 417–31.

Whitman, D. Hell in high water: Hurricane Floyd leaves behind an environmental nightmare. *U.S. News & World Report*, New York, October 4, 1999, p. 22.

Whittaker, R. *Communities and ecosystems*. New York: Macmillan, 1975.

Winslow, E. *The evolution and significance of the modern public health campaign*. London: Oxford University Press, 1984.

Woodhouse, W., and Knutson, P. Atlantic coast marshes. In Lewis, R., ed. *Creation and restoration of coastal plan communities*. Boca Raton, Fla.: CRC Press, 1982.

Woodwell, G., Wurster, C., and Isaacson, P. DDT residues in an east coast estuary: a case of biological concentration of a persistent insecticide. *Science* 156 (1967): 821–24.

World Commission on Environment and Development. *Our common future*. Oxford: Oxford University Press, 1987.

Yonge, C. The biology of coral reefs. *Advances in Marine Biology* 1 (1963): 209–60.

Zingmark, R. *An annotated checklist of the biota of the coastal zone of South Carolina*. Columbia: University of South Carolina Press, 1978.

Index

Page numbers in *italic* followed by *f* indicate figures; page numbers in *italic* followed by *t* indicate tabular material.